Alternative Futures for Digital Infrastructure

Insights and Considerations for the Department of Defense

JULIA BRACKUP, SARAH HARTING, DANIEL GONZALES, BRANDON CORBIN

Prepared for the Office of the Secretary of Defense
Approved for public release; distribution is unlimited

NATIONAL DEFENSE RESEARCH INSTITUTE

For more information on this publication, visit **www.rand.org/t/RRA1859-1**.

About RAND

The RAND Corporation is a research organization that develops solutions to public policy challenges to help make communities throughout the world safer and more secure, healthier and more prosperous. RAND is nonprofit, nonpartisan, and committed to the public interest. To learn more about RAND, visit www.rand.org.

Research Integrity

Our mission to help improve policy and decisionmaking through research and analysis is enabled through our core values of quality and objectivity and our unwavering commitment to the highest level of integrity and ethical behavior. To help ensure our research and analysis are rigorous, objective, and nonpartisan, we subject our research publications to a robust and exacting quality-assurance process; avoid both the appearance and reality of financial and other conflicts of interest through staff training, project screening, and a policy of mandatory disclosure; and pursue transparency in our research engagements through our commitment to the open publication of our research findings and recommendations, disclosure of the source of funding of published research, and policies to ensure intellectual independence. For more information, visit www.rand.org/about/research-integrity.

RAND's publications do not necessarily reflect the opinions of its research clients and sponsors.

Published by the RAND Corporation, Santa Monica, Calif.
© 2023 RAND Corporation
RAND® is a registered trademark.

Library of Congress Cataloging-in-Publication Data is available for this publication.

ISBN: 978-1-9774-1189-1

Cover: Adapted from images by Anson_iStock/Getty Images and Norcoon/Getty Images

About This Report

A competition for digital infrastructure (DI) is underway between the United States and China, which has implications for military forces and operations that rely on this infrastructure in competition and conflict. However, competition for DI remains largely understudied in a comprehensive way. This project intends to contribute to the broader understanding of DI by projecting forward to 2050. Several questions underpin this analysis regarding how DI might evolve, what it could look like in 2050, and what the DI future could mean from a military standpoint. This project is part of a multiyear research effort that defines DI, characterizes the competition underway, identifies key factors shaping outcomes, and assesses the potential implications for the U.S. Department of Defense (DoD).

This report summarizes an alternative futures analysis and assesses the military implications of these futures. This report builds on previous RAND Corporation research that characterized and defined DI.

This research should be of interest to defense analysts, strategists, technologists, and policymakers within DoD and the intelligence community assessing the impact of digital technologies and DI on competition and conflict between the United States and China.

RAND National Security Research Division

This research was sponsored by the U.S. Department of Defense and conducted within the Acquisition and Technology Policy (ATP) Program of the RAND National Security Research Division (NSRD), which operates the National Defense Research Institute (NDRI), a federally funded research and development center sponsored by the Office of the Secretary of Defense, the Joint Staff, the Unified Combatant Commands, the Navy, the Marine Corps, the defense agencies, and the defense intelligence enterprise.

For more information on the RAND ATP Program, see www.rand.org/nsrd/atp or contact the director (contact information is provided on the webpage).

Acknowledgments

The authors thank the RAND ATP management team—Chris Mouton and Yun Kang—for their guidance and support. We also thank Jarrett Caitlin, who was an invaluable member of our research team and whose early contributions helped shape the analytical products reflected in this report. The analysis contained in this report also benefited from the insights and expertise of several RAND research colleagues—Joel Predd, Michael Mazarr, Howard Shatz, Cristina Garafola, Bonnie Triezenberg, and Aaron Frank—who challenged our analytical thinking in ways that shed light on complex issues that strengthened our overall understanding of the DI. We also benefited from engagement with several researchers and practitioners outside RAND whose efforts and insights informed our thinking and the alternative futures we developed. The external engagement was conducted on a not-for-attribution basis, but we acknowledge and appreciate their willingness to expand our thinking and contribute to our analytical efforts. Finally, we thank Timothy Heath of RAND and Jim Lewis of the Center for Strategic and International Studies for their careful and constructive reviews of our report. Our work is stronger because of their insights and feedback. Any errors contained in this report are ours.

Summary

Previous RAND Corporation research defined *digital infrastructure* (DI)—a network of networks consisting of hardware and software elements that digitally process, store, and transmit data—and identified a competition underway between the United States and China for its ownership, access, and control (OAC).[1] This research highlighted the strategic importance of DI and how the degree to which a country has OAC over the DI can confer power and influence at scale, particularly in the increasingly hyperconnected world. As recently argued by two former senior U.S. leaders, "Data is the oil of the 21st century, the indispensable resource that will fuel artificial-intelligence algorithms, economic strength and national power."[2] To be sure, DI is at the heart of this competition for data and the power and influence it can yield. Additionally, this research identified important linkages between DI and military advantage. However, DI and the ongoing competition for DI remain relatively understudied in a comprehensive way. This report builds on previous research to further explore how DI might evolve and understand what impact that future DI landscape will have on military competition and conflict.

In this effort, we analyzed and identified issues and variables shaping DI outcomes. We also developed a baseline description of current DI technologies and described their status and trends. We used that analytical foundation to develop 2050 DI futures: one baseline future that depicts a straight-line projection of current trends and three alternative futures that explore novel but plausible paths. We then assessed the military implications of these futures.

We used an alternative futures approach to inform this effort. The development of alternative futures is a structured analytic approach for developing plausible futures (i.e., worlds) that differ significantly from straight-line projections of the present situation. This analytic approach serves as a tool for providing rigorous analysis to inform an otherwise uncertain future strategic landscape. An alternative futures analysis also offers insight into the impact and interaction of variables shaping strategic choices and outcomes.

Military Implications

In our analysis, we found that the global DI footprint might affect how forces operate, where forces operate, and when forces operate. First, the futures suggest that U.S. and Chinese forces might conduct military operations differently depending on the DI landscape in the area of

---

[1] DI has several components: submarine cables, wireless networks and infrastructure, terrestrial networks and infrastructure, and satellites and terminals. The DI also has two elements foundational to DI building blocks: microchips and standards. See Chapter 2 for a more detailed discussion of each DI component and foundational element. For more background on the DI, see Julia Brackup, Sarah Harting, and Daniel Gonzales, *Digital Infrastructure and Digital Presence: A Framework for Assessing the Impact on Future Military Competition and Conflict*, RAND Corporation, RR-A877-1, 2022.

[2] Matt Pottinger and David Feith, "The Most Powerful Data Broker in the World Is Winning the War Against the U.S.," *New York Times*, November 30, 2021.

operations and the types of missions. For example, power projection and longer-duration overseas presence become more difficult in any future where China has DI dominance.

Furthermore, the futures indicate that DI footprints appear to have a significant impact on military posture. Network boundaries introduce new types of borders between actors; most futures result in regional blocs determined by DI borders. The United States may choose not to have a permanent or rotational presence in territories that use China's DI given concerns over compromised information and communications. The United States can still generate combat power from the homeland and overseas, but that power generation might depend on the global DI laydown. For example, lacking an overseas permanent or rotational presence may affect how quickly the United States can flow forces to overseas locations. Sustained overseas operations will also likely look very different depending on the DI landscape. Additionally, the type of relationship with allies and partners may be limited based on DI laydown (e.g., weapon interoperability will likely be limited in more-decoupled futures; the ability to share information may be limited). As a result, the overall DI landscape could create more bifurcation between who the United States coordinates with militarily and who the United States cannot work with because of DI limitations (exacerbated by political and policy factors).

DI futures may also create a demand for new mission sets, new capabilities, and perhaps a new organization of forces. The futures show that DI could become like jet fuel in some decoupled futures—something forces need to take with them when operating overseas. For example, a new mission could be to transport and operate a mobile U.S. DI. The futures also suggest that intelligence and cyber activities may be challenged depending on the degree of U.S.-China DI interdependence or decoupling. Other mission sets or capabilities could also emerge in response to the lack of traditional cyber accesses and intelligence operations in an interconnected digital world. In a decoupled world, countries across the globe will not be interconnected like today, particularly between the United States and China. This will make cyber and intelligence operations inherently more difficult, introducing the potential need for new missions and capabilities.

Insights on Digital Infrastructure

DI appears significant for future military operations and long-term strategic competition both as an enabler and as a capability. The global DI footprint, and how it evolves, may affect how forces operate, where forces operate, and when forces operate. By doing so, DI could confer key military advantages or disadvantages for different actors and shape outcomes for competition and conflict. We have provided an initial set of insights gleaned from this analysis, with a caveat noting the speculative nature of these observations. We included them to contribute to the broader understanding of DI competition and inform future research on this topic.

Structural factors will shape the DI landscape and, ultimately, its footprint. The science, technology, engineering, and mathematics workforce (human capital more generally); economic performance and policy; industrial capacity; government approach to DI; and relationships with allies and partners all underpin what DI will look like in the future. Therefore, these factors offer important areas of focus when thinking about how to shape the future of DI in ways favorable to the United States.

DI, and the degree to which a country has OAC over DI, could serve as an enabler or a capability for military operations. Military operations may either rely on DI to execute missions (e.g., secure communications) or leverage the DI as the means to create decisive effects (e.g., a presence) or to gain access necessary to deliver an effect.

For the United States, DI appears most significant for competition activities and setting the conditions for conflict, if it were to emerge. As a result, DI OAC matters greatly for competition, and activities to determine DI OAC in competition will ultimately dictate the DI footprint for conflict. For conflict, the ability to conduct kinetic operations will likely continue regardless of DI OAC, assuming the United States has methods for basic communications and transmission of data.

Ceding DI OAC to an adversary could potentially provide an asymmetric means to erode U.S. military advantage over the long term. While the U.S. military can likely conduct military operations without DI OAC (i.e., the United States could likely conduct strike missions without control of the DI), access to information and systems could be leveraged for an advantage in degrading or exploiting U.S. capabilities, operations, and other functions, such as planning.

Contents

Figures and Tables

Figures

Tables

CHAPTER 1

Introduction

A competition for digital infrastructure (DI) is underway between the United States and China, which affects military forces and operations that rely on this infrastructure in competition and conflict. However, competition for DI remains largely understudied in a comprehensive way. How might DI evolve? What could DI look like in 2050? What does advantage and disadvantage look like across different futures? And what are the implications of these futures for military competition and conflict? Our research effort sought to shed light on these questions.

Background

Previous RAND Corporation research defined and characterized *DI*—a network of networks consisting of hardware and software elements that digitally process, store, and transmit data—and their associated foundational elements (microchips, standards).[1] Through this research, we identified a competition for DI underway between the United States and China, with concerted efforts by both countries to leverage DI and its foundational elements for military applications. We found that digital dual-use technologies associated with DI may serve as a force multiplier for military posture and presence and create key intelligence advantages. Furthermore, we found that both the United States and China rely on DI to support military forces and are using it to expand national power and influence globally. We also found that a country's ownership, access, and control (OAC) of DI and the data associated with it could have an impact on U.S.-China competition and conflict given that traditional elements of power (military, economic, information, demographic) are being shaped increasingly by DI.

DI has evolved over many decades. After World War II, the United States was a leader in all aspects of DI, but over time, DI leadership has become more diffuse and competitive. Today, the United States and China remain major leaders in DI, aiming to shape it in ways that align with their long-term strategic interests. However, U.S. and Chinese goals and

---

[1] The DI has several components: submarine cables, wireless networks and infrastructure, terrestrial networks and infrastructure, and satellites and terminals. For more background on the DI, see Julia Brackup, Sarah Harting, and Daniel Gonzales, *Digital Infrastructure and Digital Presence: A Framework for Assessing the Impact on Future Military Competition and Conflict*, RAND Corporation, RR-A877-1, 2022.

approaches for DI differ. The future of DI remains uncertain, but preliminary research indicates that it will be vital to future warfare and strategic competition.[2]

Challenges persist in understanding how the strategic competition for DI might play out and its implications for military competition and conflict.[3] While many factors and variables will shape outcomes associated with DI competition, identifying these factors and choosing which ones to emphasize for further analysis remains difficult. The lack of historical precedent for DI competition exacerbates this point. As a result, identifying and understanding which variables and which interactions among variables might shape DI outcomes requires analytic methods that allow for the interaction of many factors to discern insights. Alternative futures analysis offered an attractive option given its use of variables to develop different outcomes. We leveraged previous RAND research on DI to inform status and trends and then used informed judgment and context to develop and assess potential DI futures. We sought to identify and understand the broad set of variables—and their interactions—associated with these futures to inform potential areas of future analysis and to offer insights into where the United States can create military advantage. The analysis presented in this report should not be viewed as definitive but rather as an informed approach to think about the U.S.-China competition for DI.

Objective

The objective of our study was to generate and evaluate a set of plausible 2050 alternative futures for the evolution and control of the global DI and assess their potential implications for U.S.-China military competition and conflict. The alternative futures intended to be novel but plausible and challenge assumptions and straight-line projections. The alternative futures also intended to inform our understanding and identification of factors shaping DI outcomes.

Approach

We leveraged an alternative futures approach to inform this research. The development of alternative futures is a structured analytic approach originating from the U.S. intelligence community for developing plausible futures (i.e., worlds) that differ significantly from straight-line projections of the present situation. This analytic approach serves as a tool for

---

[2] Brackup, Harting, and Gonzales, 2022.

[3] We acknowledge that this is a long-standing analytic challenge for strategic choices and strategic questions writ large. As Barry Watts has noted, "Choosing analytic criteria for making strategic choices or judging historical outcomes is a recurring, if not universal, problem" (Barry Watts, *Analytic Criteria for Judgments, Decisions, and Assessments*, Center for Strategic and Budgetary Assessments, 2017, p. 1).

providing rigorous analysis to inform an otherwise uncertain future strategic landscape.[4] An alternative futures analysis also offers insight into the impact and interaction of variables shaping strategic choices and outcomes.

For the purposes of this report, we define alternative futures and related terms as follows:

- *issues*: the broadest category of developments, questions, knowns, unknowns, etc., that could influence one or more alternative futures (e.g., dependency on other countries for DI)
- *variable*: a critical factor or set of factors that will have a defining impact on one or more alternative future (e.g., U.S. ownership of submarine cables)
- *alternative future*: a plausible future (i.e., world) that differs significantly from straight-line projections of the present situation
- *scenario*: a plausible path that could lead to or from the future that takes into account specific variables (e.g., China's push to control the Taiwan Semiconductor Corporation [TSMC] and Taiwan).

Our analytic approach for this study was divided into several steps. First, we identified issues across the political, military, economic, social, and information dimensions that will have a defining impact on outcomes for DI and its foundational elements. We provided an analytical basis and background information to justify the identification of these issues, informed by engagement with a community of interest (COI) consisting of RAND and external regional and functional subject-matter experts (SMEs). Next, and building on previous DI research,[5] we developed a baseline description of current DI technologies and described their status and trends. We used this DI baseline and the issues identified in the first step to develop a straight-line projection of a DI baseline future for the 2050 time frame. We then developed plausible alternative DI futures for the 2050 time frame that varied from straight-line projections and were based on different interactions or relationships between the issues and variables.

The development and refinement of the DI baseline future and the alternative futures was informed by further engagement with a COI of SMEs in a structured analytic workshop. We used the alternative futures to assess their military implications. The assessment of military

---

[4] RAND researchers have leveraged an alternative futures analysis to inform long-term strategic planning efforts for U.S. Strategic Command, the Department of the Air Force, and the Office of the Secretary of Defense for Research and Engineering, among others. For more background on developing alternative futures and related structured analytic techniques, see Richards J. Heuer, Jr., and Randolph H. Pherson, *Structured Analytic Techniques for Intelligence Analysis*, CQ Press, 2011; Robert J. Lempert, Steven W. Popper, and Steven C. Bankes, *Shaping the Next One Hundred Years: New Methods for Quantitative, Long-Term Policy Analysis*, RAND Corporation, MR-1626-RPC, 2003; James A. Dewar, *Assumption-Based Planning: A Tool for Reducing Avoidable Surprises*, Cambridge University Press, 2002; and Richard E. Neustadt and Ernest R. May, *Thinking in Time: The Uses of History for Decision-Makers*, The Free Press, 1986.

[5] Brackup, Harting, and Gonzales, 2022.

implications was informed by a literature review of U.S. Department of Defense (DoD) and the People's Liberation Army (PLA) efforts to leverage DI now and in the future for military operations. We used that literature review to identify core assumptions associated with U.S. and Chinese military reliance on DI and then assessed the implications of the DI futures given those assumptions.

Organization of This Report

This report is organized into several major chapters. Chapter 2 describes the competition for DI and provides a detailed discussion of DI technologies and their status and trends. Chapter 3 describes a DI baseline future and three 2050 alternative futures for DI, along with a discussion of the key assumptions and variables in our analysis. Chapter 4 discusses the military implications of these futures. Chapter 5 concludes with a set of insights and considerations on DI for DoD. The appendix provides the matrix used for each future to illustrate the variables and settings in greater detail.

CHAPTER 2

Understanding the Future of Digital Infrastructure

We developed a conceptual approach for understanding the competition for DI. The approach involved identifying key elements of the DI competition that can be applied to each DI technological building block to frame the construction of several alternative futures. We intentionally focused the alternative futures on the technology development competition, narrowly defined below. While our broader definition and understanding of the DI competition described in earlier studies includes geopolitical and military dimensions, we excluded these elements in constructing DI alternative futures.[1] In doing so, we aimed to understand the trajectory of the technology landscape—to include U.S-Chinese relative DI ownership—and then in subsequent chapters analyze geopolitical and military considerations. In this chapter, we briefly walk through how we conceptualized the DI competition to inform both the futures and the implications discussed in Chapter 4.

The Competition for Digital Infrastructure

Elements of the Competition

Collectively, the competition elements illustrate the DI global technological landscape. We begin with the early stages of technology development.

Technology Development

We use *technology development* to mean the DI research and development (R&D) performed by innovators worldwide, including DI system and software design leaders and the manufacturing leaders of key DI components, including microchips. Therefore, the variables we identified as significant to developing DI technology are as follows:

- science, technology, engineering, and mathematics (STEM) workforce
- DI R&D groups
- DI standards-setting organizations
- industrial base

---

[1] Brackup, Harting, and Gonzales, 2022.

- government approach
- economic policy
- economic performance.

We conducted a literature review, engaged SMEs, and held a workshop to inform the identification of these variables. First, consensus exists that to develop highly complex and cutting-edge DI technology, a country must have the requisite talent.[2]

Second, healthy R&D organizations and a sound industrial base—including sufficient talent, intellectual property (IP), and industrial capacity, such as organizational capacity and capital—are essential for the development of DI.[3] In particular, given that current and future DI will continue to be developed largely by private sector companies, multinational industrial bases will likely continue to be important for DI.

Third, how governments approach DI also affects how companies, and potentially the public sector, develop and invest in DI technology.[4] For example, a systematic government approach might include large and strategic investments in the development of cutting-edge DI technology, or a siloed or highly regulated government approach, that slows or hampers DI innovation. This could discourage investment in R&D or implementation of new DI networks.[5] On the other hand, the lack of a government approach, especially in R&D decision-making, could also lead to more rapid and effective innovation by the private sector because of healthy competition among companies.

Last, we identified two variables tied to the economy: economic performance and economic policy. Today, DI technology stands out as both resource intensive and capital intensive, requiring long-term investments and consistent streams of capital. Therefore, having the necessary economic foundation to support such financial demands will likely remain critical to how a country develops its DI technology. After the development of DI occurs, countries then adopt and use it to varying degrees, creating what we term the *global DI footprint*.

---

[2] Alan Zilberman and Lindsey Ice, "Why Computer Occupations Are Behind Strong STEM Employment Growth in the 2019–29 Decade," *Beyond the Numbers: Employment & Unemployment*, Vol. 10, No. 1, January 2021; Becky Frankiewicz and Tomas Chamorro-Premuzic, "Digital Transformation Is About Talent, Not Technology," *Harvard Business Review*, May 6, 2020; Elliot Silverberg and Eleanor Hughes, "Semiconductors: The Skills Shortage," The Lowy Institute, September 15, 2021; Stephanie Yang, "Chip Makers Contend for Talent as Industry Faces Labor Shortage," *Wall Street Journal*, January 2, 2022.

[3] James Andrew Lewis, *Mapping the National Security Industrial Base: Policy Shaping Issues*, Center for Strategic and International Studies, May 2021.

[4] Jon Bateman, *U.S.-China Technological "Decoupling": A Strategy and Policy Framework*, Carnegie Endowment for International Peace, 2022.

[5] A key regulatory hurdle that has threatened to slow the rollout of fifth generation (5G) and low earth orbit (LEO) satellite networks is the allocation of new wireless spectrum for these networks by regulatory authorities in various countries.

International Use and Adoption of Digital Infrastructure

How various countries, companies, and populations adopt and use DI technology creates a global laydown, or footprint, of DI. These variables inform how different DI futures might evolve and remain speculative in nature. Additionally, we use such variables in an attempt to understand what factors might influence how other actors perceive DI technology and their decisions to choose one technology provider over another. This competition element naturally stems from the development of the technology. The variables underpinning the international use and adoption of DI are as follows:

- overseas DI strategy
- relationships with allies and partners.

First, a country's strategy for selling DI overseas will likely affect the international use and adoption of DI. We envisioned this variable across a spectrum of settings, ranging from a comprehensive national-level DI overseas strategy to a nonexistent approach to overseas DI sales. Inherent to this variable is the contrast between China's Digital Silk Road (DSR) initiatives and those of the United States.[6] The U.S. government does not have an initiative akin to China's DSR currently. However, that could change in the future. We note here that the lack of a U.S. overseas DI strategy might not be disadvantageous but rather simply an asymmetry in how either country approaches the technology from a top-down perspective.

Second, a country's relationship with other countries will likely affect how DI becomes adopted globally. For example, while the United States does not currently have a systematic top-down approach like the Chinese DSR for DI, it does have relationships with other countries that can be leveraged to influence how these countries adopt and use DI. We measured this variable across a spectrum from strong and robust relationships to weak and strained relations.

Direct Dependencies on Other Actors

To further elucidate how the competition might evolve, we added an element on the degree to which China and the United States have access to alternative DI technology if they either lack ownership, access, control (OAC) or do not have enough domestic capacity to self-sufficiently access the technology. The two variables we noted as critical to this element are as follows:

- overall OAC of DI
- access to alternatives.

For overall OAC of DI, we used a spectrum ranging from full monopoly for an actor to fully reliant on others. For access to alternatives, the spectrum of options ranges from access to trusted alternatives to no alternatives. Whether the United States or China will have little

---

[6] Jonathan E. Hillman, *The Digital Silk Road: China's Quest to Wire the World and Win the Future*, Harper Collins Publishers, 2021; Joshua Kurlantzick, "Assessing China's Digital Silk Road Initiative: A Transformative Approach to Technology Financing or a Danger to Freedoms?" Council on Foreign Relations blog, December 18, 2020.

to no dependency on others will be determined by their degree of global DI OAC. Thus, in futures with a dominant China, it becomes important whether the United States can access DI from trusted partners or whether it must use a Chinese-made product for parts of the DI. Similarly, if China cannot develop DI domestically, who might it turn to for alternatives? Much overlap and interaction exist between the three competition elements. For example, if the United States does not have OAC of wireless DI technology and must access it from another country, its relationships with allies and partners (and their roles with respect to DI) become particularly important.

Each element of the competition consists of a series of variables that can be thought of as dials with different settings. We turned the dials to create different settings that lead to varying future environments for the DI. Table 2.1 presents these elements, their respective variables, and the settings for each.

TABLE 2.1

Key Competition Variables and Settings

Element	Variable	Setting A	Setting B	Setting C	Setting D
Tech development	STEM workforce	Strong	Sufficient	Declining	Insufficient
	Industrial base	Self-sufficient; DI technology protected, secure	Some reliance on untrusted 3rd parties; large domestic capacity	Mostly reliant on 3rd parties; limited domestic capacity	Reliance on untrusted 3rd parties
	Government approach	Systematic/ centralized	Siloed	Nonexistent by choice	Diminished government role/ authority
	Economic policy	Laissez faire/ free market	Centralized/ directed	Mixed—some markets closed	Open to allies, closed to adversaries
	Economic performance	Global leader	Growing	Declining	Poor
International use and adoption	Overseas DI strategy	Comprehensive national-level DI strategy and resources	Some offensive (funding), some defensive measures	Defensive only policy (no use of adversary infrastructure)	Nonexistent
	Relationships with allies and partners	Strong	Strained	Limited	Weak
	Overseas OAC of DI	Global	Many regions	Own and allied countries	None
Access to alternatives	Overall OAC of DI	Dominant/ monopoly	Diffuse/ competitive	Limited	None; reliant on others
	Access to alternatives	Yes; trusted	Yes; untrustworthy	Limited	None

This matrix informed and structured our development of alternative futures by stressing different assumptions, variables, and settings to create various outcomes for the DI in 2050. Chapter 3 walks through these futures and the variables emphasized for each. The appendix presents these matrices for each future as well.

Elements of Digital Infrastructure

How the individual elements of DI—or building blocks—interact with the variables in Table 2.1 will collectively shape how the DI evolves. We explore each building block in the next section to think through various ways the DI might look in the future, with each element offering a different insight. For instance, the United States might have dominance in space and submarine cables but cede advantage to China in terrestrial and cellular networks. What does advantage look like in this future DI landscape? As noted earlier, how a country develops the technology, how the world adopts and uses it, and the degree to which China and the United States have OAC will likely differ for each part of the DI. Consistent with previous RAND work characterizing the DI,[7] the building blocks are as follows:

- terrestrial networks
- cellular networks
- space networks
- submarine cable networks.

We also included two elements foundational to DI building blocks.

- microchips
- standards.

In Table 2.2, we provide a brief definition and description of each building block.

TABLE 2.2
Definitions of Digital Infrastructure Building Blocks

DI Building Block	Definition
Terrestrial	Wireline and fiber-optic networks and the information infrastructure these networks link, including data centers, cloud computing infrastructure, personal computers (PCs), and other user devices
Cellular	Wireless network infrastructure, including mobile devices
Space	Satellites and satellite terminals
Submarine cable	High-speed, high-capacity undersea telecommunications links

SOURCE: Features information from Brackup, Harting, and Gonzales, 2022.

---

[7] See Brackup, Harting, and Gonzales (2022) for more details on each building block.

In the next section, we walk through each element of the DI in greater detail to understand how and where an actor can develop OAC and what advantage in DI might look like in different futures.

Digital Infrastructure Technologies

After the invention of the telephone and radio in the 19th century, humans expanded their ability to communicate over longer distances. Initially, this communications infrastructure was based on analog technologies that could transmit only limited amounts of information. After the invention of digital communications by DoD in the Advanced Research Projects Agency Network (ARPANET) program in 1969, the capabilities of communications networks expanded dramatically. DI now spans the globe and includes four types of networks: terrestrial, wireless, submarine cable, and satellite.[8]

Below we describe how these networks are organized, the key technologies they rely on, and key vendors in the supply chains for each network segment. In our earlier work, we introduced the concept of DI OAC.[9] In the discussion below, we delve into the supply chain for each network and examine how network equipment vendors can potentially gain OAC of networks they support.

Terrestrial Networks

Network Description

Terrestrial networks are used to link data centers, cloud computing infrastructure, PCs, servers, and a wide range of consumer and industrial devices. Terrestrial networks are also used to access, monitor, and control critical infrastructure, such as water utilities, oil and gas refineries, power plants, and the electric power grid. Modern terrestrial networks are digital; they digitally process, store, and transmit data. Modern terrestrial networks are based on internet communications protocol standards, whereas legacy terrestrial networks are based on older Signaling System 7 (SS7) standards, as well as other technical standards that apply to lower levels of the "technology stack."[10] Both sets of technical standards enable global connectivity between terrestrial networks owned and operated by different telecommunications companies or internet service providers (ISPs). SS7 standards were originally developed to support voice and low data rate communications between user devices with assigned telephone numbers. In SS7 networks, user devices (e.g., traditional telephones) are connected to a central office or local exchange switch using wires or wirelines. Legacy terrestrial networks use copper wiring (i.e., twisted pair cables).

---

[8]   Brackup, Harting, and Gonzales, 2022.

[9]   Brackup, Harting, and Gonzales, 2022.

[10]   Andrew Tannenbaum, *Computer Networks*, 4th ed., Prentiss Hall, 2003.

The invention of the internet enabled users to connect to the network with a wide range of communications devices, including PCs, servers, smart televisions, and mobile devices. The rapid development of internet technologies supported healthy competition between internet infrastructure providers, which led telecommunications companies to converge their networks into a single system based on the core Internet Protocol. Today, the internet provides the underlying foundation for terrestrial networks. Modern terrestrial networks no longer use twisted pair cabling and instead use higher-capacity coaxial cables or fiber-optic links. One of the key innovations of the internet involved decentralizing message switching and decomposing messages into data packets. The use of internet routers and switches makes the internet much more flexible, fault tolerant, and adaptive. New communications standards were developed to support high-capacity communications in local area networks (LANs), wide area networks (WANs), and metropolitan area networks (MANs). Ethernet transceivers are typically used in LANs. PCs, printers, and servers use ethernet transceivers to connect to LANs. High-capacity WANs and MANs can use a variety of transceivers and, increasingly, use optical transceivers.

Fiber-optic subnetworks in terrestrial networks support very high-capacity communications links and form the backbone of large national-level terrestrial networks. Data are transmitted by optical transceivers as light pulses in optical fiber. Long-distance fiber-optic links in terrestrial networks might use optical repeaters to ensure that data are not lost over long-distance fiber links.

Network Components and Supply Chains

Key components of terrestrial networks are listed in the top row shown in Figure 2.1. U.S. companies pioneered the development of SS7 networks and the internet, their associated components, and technical standards.[11] U.S. companies, such as Cisco Systems, Arista, HPE, and others have a large share of the internet equipment market, as indicated in the figure, although Chinese companies, such as Huawei and H3C, also have a significant market share. Terrestrial components are color coded in Figure 2.1 to indicate whether the United States has an advantage or disadvantage in specific component sectors. Figure 2.1 indicates that companies from a single country do not dominate supply chains for any components of terrestrial networks and that U.S. companies have a strong position in three of the five sectors. Consequently, U.S. terrestrial network providers, such as AT&T and Verizon, do not depend on China for any key components of their terrestrial networks.

Short-range wireless networks are an extension of terrestrial networks and use Wi-Fi routers based on the 802.x series of technical standards from the Institute of Electrical and Electronics Engineers. Wi-Fi routers enable communications to dozens of devices simultaneously. Wi-Fi routers were first made in the United States by U.S. companies. The market share of U.S. companies in this sector has fallen as companies such as Cisco Systems and Apple have left the market. Consumer-grade Wi-Fi routers are now inexpensive commodity

---

[11] Brackup, Harting, and Gonzalez, 2022.

FIGURE 2.1

Terrestrial Networks and Supply Chains

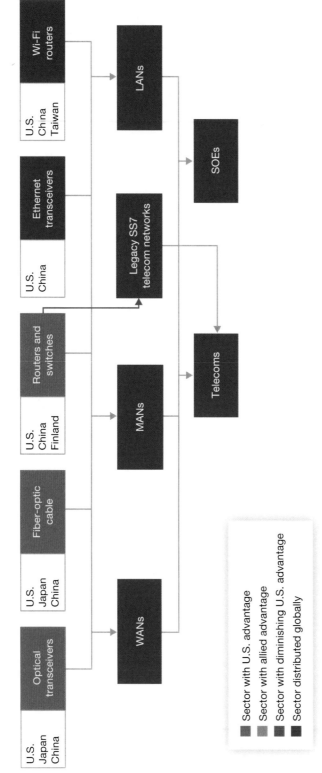

SOURCES: Authors' analysis of information from Absolute Reports, "Wireless Router Market Size by 2022–2028 Key Players, Regional Segmentation, Types, Applications, Growth, Shares, Revenue, Opportunities, Challenges, Drivers, Trends," press release, April 29, 2022; Research and Markets, "Global Optical Transceiver Market (2022 to 2027)—Featuring Amphenol, Broadcom and Infinera Among Others," press release, June 30, 2022; International Data Corporation, "IDC's Worldwide Quarterly Ethernet Switch and Router Trackers Show Strong Growth in Fourth Quarter of 2021," press release, March 10, 2022; Reports and Data, *High-Speed Optical Transceiver Market*, 2022; Expert Market Research, "Global Ethernet Adapter Market Report and Forecast 2022–2027," webpage, undated: Mike Robuck, "Report: High-Speed Data Center Ethernet Adapter Market Tops $1B for the First Time," Fierce Telecom, May 20, 2020; Telecom Lead, "Cisco Replaces Huawei in Service Provider Router and Switch Market," March 11, 2022; Trish Novico, "Five Largest Fiber Optic Companies in the World," *Insider Monkey*, December 3, 2020; Ken Wieland, "Router/Switch Market Back to Pre-Pandemic Levels—Dell'Oro," Light Reading, June 1, 2021.

NOTE: SOE = state-owned enterprise; telecom = telecommunications; telecoms = telecommunications providers.

products. A few U.S. companies, such as Netgear and Belkin, still make routers, but most are now made by Chinese and Taiwanese companies that specialize in low-cost manufacturing.

Terrestrial networks are owned and operated by (1) private companies in the United States, Europe, and other countries or (2) by SOEs in China and other countries, as shown in the bottom row of Figure 2.1. The telecommunications industry, which includes terrestrial network providers, is highly regulated in many countries. Government regulators carefully review proposed changes in ownership and the actions of telecommunications providers. In some countries, although not in the United States, foreign companies are not permitted to own a majority stake in a telecommunications provider. For these reasons, it is reasonable to assume that in most countries terrestrial network providers are trustworthy owners of their networks and will not support espionage operations of a foreign country. However, it is also possible that a terrestrial network could become compromised because it relies on untrustworthy components. The data that inform Figure 2.1 suggest that this should not be a concern in the United States because there are trusted sources for all major components of terrestrial networks.

Wireless Networks

Network Description

The capability of wireless cellular networks has grown significantly from their earliest versions, to when wireless telecommunications providers rolled out 5G cellular networks around the world in 2022. 5G networks will provide higher-capacity links, at lower latency, to a larger number of end user devices than previous wireless networks were capable of.[12] Figure 2.2 shows key components of a 5G network highlighted in red: core network servers, edge servers, and the base stations and antennas of the radio access network (RAN). In contrast to earlier cellular networks, 5G networks are increasingly software defined so their core functions can be provided and hosted in cloud-based infrastructure. Software-defined cloud-based 5G core servers can be scaled up and down in response to changing user demand conditions, enabling cellular network providers to save money on hardware.[13] 5G networks will provide connectivity to a wider range of end user devices, such as cell phones and tablets, Internet of Things (IoT) devices, and connected autonomous vehicles (see Figure 2.2).

The three parts of 5G networks—the core, RAN, and edge cloud—are connected by a backhaul network that can be provided by terrestrial networks or special purpose high-capacity wireless links, as shown in the figure. The 5G network will be monitored and con-

---

[12] Timothy M. Bonds, James Bonomo, Daniel Gonzales, C. Richard Neu, Samuel Absher, Edward Parker, Spencer Pfeifer, Jennifer Brookes, Julia Brackup, Jordan Wilcox, David R. Frelinger, and Anita Szafran, *America's 5G Era: Gaining Competitive Advantages While Securing the Country and Its People*, RAND Corporation, PE-A435-1, 2021.

[13] Cloud-based dependence of 5G infrastructure could, however, also introduce new cyber vulnerabilities and access points for espionage and intelligence collection operations by nation states and untrustworthy network equipment suppliers.

FIGURE 2.2

Overview of 5G Wireless Networks

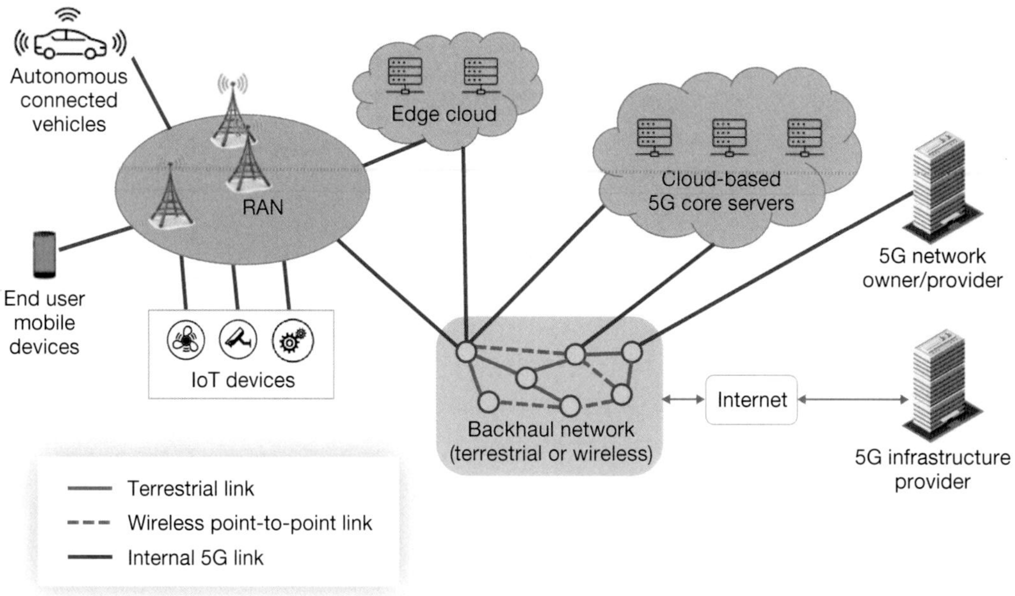

SOURCE: Adapted from Daniel Gonzales, Julia Brackup, Spencer Pfiefer, and Timothy Bonds, *Securing 5G: A Way Forward in the U.S. and China Security Competition*, RAND Corporation, RR-A435-4, 2022.

trolled by the wireless network owner or provider. Typically, the 5G network provider retains exclusive control of the network. However, the 5G equipment provider will likely have access to the network so it can supply software updates and troubleshoot the network when network anomalies or troubles occur.

Network Components and Supply Chain

Figure 2.3 shows the key components of a wireless network, the countries that are major players in these equipment markets, the types of companies that own and operate 5G networks, and our assessment of the U.S. position in each of the equipment sectors or markets. Color coded similar to Figure 2.1, Figure 2.3 shows that U.S. telecommunications providers (telecoms) have access to a trustworthy supply chain for all types of 5G network components. However, U.S. firms lead in only one area, routers and switches, and the United States must rely on foreign suppliers for 5G wireless base stations and core network servers. Fortunately, the United States can obtain trustworthy equipment from suppliers in Europe for 5G core networks and base stations, although Huawei and ZTE remain major suppliers of these 5G components, especially to telecoms operating in Asia and many developing countries.

Figure 2.3 also shows that Chinese companies offer products in all wireless network component categories. 5G technical standards enable private companies to run 5G networks in small areas and inside facilities, such as factories or ports. 5G can potentially support industrial automation and real time control of industrial processes in chemical plants and facto-

FIGURE 2.3

Wireless Networks and Supply Chains

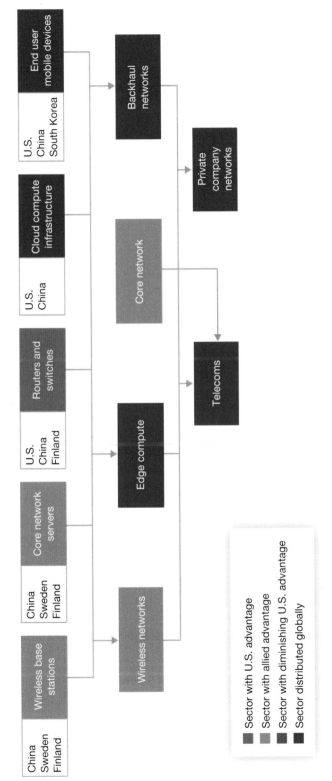

China Sweden Finland	China Sweden Finland	U.S. China Finland	U.S. China	U.S. China South Korea
Wireless base stations	Core network servers	Routers and switches	Cloud compute infrastructure	End user mobile devices

Wireless networks

Edge compute

Core network

Telecoms

Backhaul networks

Private company networks

Sector with U.S. advantage

Sector with allied advantage

Sector with diminishing U.S. advantage

Sector distributed globally

SOURCE: Based on data from Gonzales et al., 2022.

ries. Private 5G networks could also potentially be compromised by an untrustworthy 5G infrastructure provider.

Technical Standards for Wireless Networks

The technology for 5G networks took more than a decade to develop by the international technical community. All aspects of 5G networks are defined by 5G technical standards, which were jointly developed by technical experts from leading companies in the wireless communications industry. The 3rd Generation Partnership Project (3GPP) is the international standard-setting organization responsible for developing 5G, and next 6G, technical standards for cellular networks. Such technical standards have increasingly become areas of competition between not only companies but also countries and, in particular, between China and the United States. In this report, we will not delve into the details of the technical standard-setting process used by 3GPP. Here we only note that technical standards are an important foundational element for all aspects of the DI, and, therefore, they will likely remain an area of competition between the United States and China. They will also play a role in the alternative futures discussed later in this report.

Submarine Cable Networks

Network Description

Ninety-seven percent of the internet traffic transmitted between continents is carried by submarine cable networks.[14] The newest, most advanced submarine cables can carry hundreds of terabytes per second of data, using just 24 strands of fiber-optic filament. Submarine cables are strategic assets and cost hundreds of millions of dollars to build and install on the ocean floor. The ownership and landing points of submarine cables in the Pacific are closely scrutinized by the U.S. government. Recently, a cable partly owned by Google and Facebook had a terminus in the United States and its other landing site in the Philippines, and it was to have an extension that landed in Hong Kong. Later, it was discovered that one of the true owners of the cable system had ties to China. As a result, the cable extension to Hong Kong was denied by the U.S. government. The U.S. government did not provide a specific explanation for this action.[15]

In our prior research, we showed that submarine cables that land on U.S. territory are typically co-owned by private U.S. companies, such as Google or Amazon.[16] The other owners of such submarine cables will typically be private companies or telecoms based in the other country or countries with the cable landing sites. In our previous analysis, we found that

---

[14] James Glanz and Thomas Nilsen, "A Deep-Diving Sub. A Deadly Fire. And Russia's Secret Undersea Agenda," *New York Times*, April 20, 2020.

[15] U.S. Department of Justice, "Team Telecom Recommends That the FCC Deny Pacific Light Cable Network System's Hong Kong Undersea Cable Connection to the United States," press release, June 17, 2020.

[16] Brackup, Harting, and Gonzales, 2022.

submarine cables that were co-owned by Chinese companies were, for the majority of them, Chinese telecoms.[17]

Network Components and Supply Chains

Figure 2.4 shows other important elements of a submarine cable network. Parts of a cable located near the shore and on the continental shelf are usually armored to protect them from breakage or cuts from ship anchors and sharks. In deep water, cables are typically not armored and are much lighter. Many submarine cables running underneath the world's major oceans are thousands of kilometers long, and so, the light signals that carry information will attenuate over such distances and could not be deciphered at the receiving landing site unless they are periodically amplified or retransmitted. For this reason, optical repeaters are installed along the submarine cable as shown in the figure. Optical repeaters require significant amounts of power; thus, submarine cables require hundreds and sometimes thousands of kilowatts to power the string of repeaters installed in the network.

On the right side of Figure 2.4 is a landing site in one country—for example, the United States. The landing site is a collection point where a wide range of terrestrial and wireless networks connect to provide a transit point to the submarine cable so users in the United States can communicate with users in other countries connected to the cable. Major components of submarine cable networks are shown in the top row of Figure 2.5, as well as the countries that manufacture each type of component. Several components used in other types of networks are also used at cable landing sites to transmit, receive, and route information, such as digital multiplexers, photonic transceivers, routers, and switches.

A unique aspect of submarine cable networks is the ships used to install the submarine cables. These special-purpose ships carry large spools of fiber-optic cable and equipment used to lay cable on the ocean floor. Cable-laying ships can also be used to repair cables that suffer damage by raising damaged cable components to the ship where they can be repaired. Only a few companies have these types of ships and experience in cable-laying operations. In 2014, Huawei purchased a UK cable-laying company and created a new subsidiary called Huawei Marine. After concerns grew about Huawei's growing marine market share and its role in laying cables that used Huawei equipment connecting to the telecommunications networks of North Atlantic Treaty Organization (NATO) allies,[18] Huawei sold its submarine cable–laying subsidiary to another Chinese company, Hengtong, in 2019.[19] In contrast, not until 2021 did the United States have its own cable-laying ships that it could call on in wartime or crisis to repair severed cables. The discussion above that describes how a submarine

---

[17] Brackup, Harting, and Gonzales, 2022.

[18] Jeremy Page, Kate O'Keeffe, and Rob Taylor, "America's Undersea Battle with China for Control of the Global Internet Grid," *Wall Street Journal*, March 12, 2019.

[19] Stephen Hardy, "Hengtong to Buy Huawei Marine Networks," Lightwave, November 4, 2019.

FIGURE 2.4

Components of a Submarine Cable Network

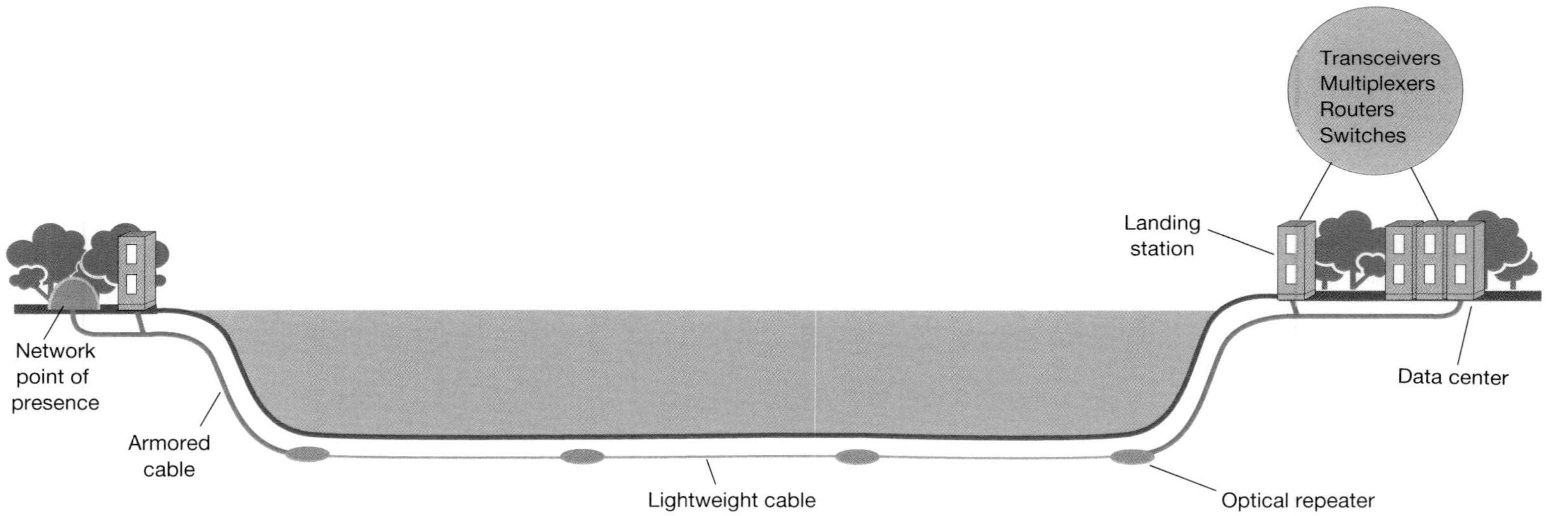

Transceivers
Multiplexers
Routers
Switches

Data center

Optical repeater

Landing
station

Lightweight cable

Network
point of
presence

Armored
cable

FIGURE 2.5

Submarine Cable Networks and Supply Chains

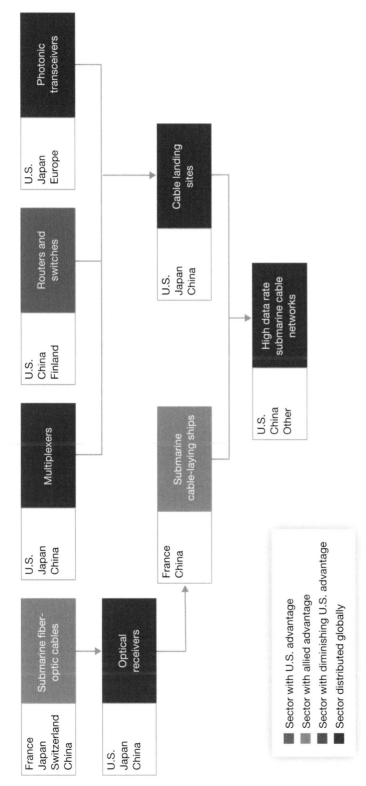

SOURCES: Authors' analysis of information from Doug Brake, "Submarine Cables: Critical Infrastructure for Global Communications," Information Technology and Innovation Foundation, April 19, 2019; Brackup, Harting, and Gonzales, 2022; "Top 5 Vendors in the Global Submarine Fiber Cable Market from 2017 to 2021," Business Wire, October 30, 2017; Wieland, 2021.

cable could be tapped during the installation process raises concerns as to whether China has used Huawei or Hengtong to tap Chinese submarine cables.

The situation with Hengtong prompted changes in U.S. policy regarding submarine cables. After some resistance within U.S. civilian government agencies, Congress directed the Department of Transportation to create a Cable Security Fleet and provided funding to ensure that two U.S.-flagged cable-laying ships will be available within 24 hours to securely repair any damaged U.S. submarine cables. Furthermore, these two cable-laying ships will be crewed only by U.S. citizens.[20]

Figure 2.5 shows that the United States is not currently at a disadvantage in the submarine cable market supply chain. U.S. companies co-own a significant percentage of the submarine cables in the Pacific and Atlantic oceans.[21] Furthermore, the United States does not depend on China for its supply chain. Nevertheless, China's growing role in the market for submarine cable installation and the increasing number of submarine cables co-owned by Chinese telecoms raise concerns over China's growing control and influence in this segment of the global DI.

Satellite Networks

Network Description

Communications satellites play an increasingly important role in the global DI. They provide communications links to stationary facilities—ones that might be far from high-capacity terrestrial or wireless networks—and mobile platforms, such as aircraft, ships, and ground vehicles. The capabilities of commercial communications satellites (COMSATs) have grown significantly over time, following a pattern similar to that seen in the other segments of DI. The latest generation of COMSATs provides more capacity and can support many more users than earlier systems.

As of August 2022, there were more than 4,800 satellites in space, and this number will continue to increase as more companies enter the market.[22] The space economy is growing at a rapid rate. The number of satellites in orbit has grown by a factor of five from 2012 to 2021. Most of the growth in space is in COMSATs, because most satellites in orbit in 2022 (58 percent of the total) are COMSATs.[23]

It is estimated that the COMSATs in orbit by 2026 will provide close to 200 terabytes per second (Tbps) of capacity.[24] Of that, 6 Tbps will be provided by satellites planned for geostation-

---

[20] Douglas R. Burnett, "Submarine Cable Security and International Law," *International Law Studies*, Vol. 97, No. 1, 2021.

[21] TeleGeography, "Submarine Cable Map," dataset, 2023.

[22] Satellite Industry Association, *State of the Satellite Industry Report*, June 2022.

[23] Satellite Industry Association, 2022.

[24] Satellite Industry Association, 2022.

ary orbit. The remainder, more than 190 Tbps, will be provided by LEO or medium earth orbit (MEO) satellites. [25] COMSAT providers are deploying new constellations in LEO where communications time delays are minimized, enabling a true extension of the internet into space around the earth for the first time. SpaceX, a U.S. company, is leading industry in deploying a large-scale COMSAT constellation into LEO. As of August 2022, SpaceX had over 3,000 Starlink COMSATs in LEO.[26] And as of June 2022, Starlink had over 400,000 subscribers worldwide.[27]

U.S. and European companies dominate the COMSAT manufacturing market. U.S. and European companies, such as Intelsat, Inmarsat, Viasat, OneWeb, and SpaceX, own the largest COMSAT constellations. Traditional COMSATs are placed in geosynchronous orbit (GEO) and appear to be stationary to users on the ground. The earliest COMSATs were placed in GEO because simpler, much less expensive, nontracking satellite communications (SATCOM) terminals could be used with them. However, even though GEO COMSATs have grown in size and capability, they can only provide so much capacity. For example, the highest-capacity COMSAT now planned for GEO will have a capacity of 1 Tbps, which is a substantial increase in capacity over older GEO COMSATs.[28] Nevertheless, its capability is dwarfed by the capacity provided by submarine cables. For example, NEC Corporation will build a new transatlantic cable for Facebook that will have a capacity of 500 Tbps.[29]

Industry has moved toward LEO COMSATs in part because of the lower communications time delays these systems offer over GEO COMSATs but also because they can provide high-capacity links using smaller, less expensive satellites that can be mass produced, enabling economies of scale not possible with GEO COMSATs. As mentioned earlier, the SpaceX Starlink constellation is composed of about 3,000 satellites. In its Federal Communications Commission filing, SpaceX estimated each Starlink satellite could offer a maximum capacity of 23 gigabytes per second (Gbps).[30] If we assume all Starlink satellites in orbit can operate at this maximum data rate, the entire Starlink constellation could provide an aggregate capacity of 69 Tbps, still a small fraction of the capacity of one new submarine cable.

---

[25] Satellite Industry Association, 2022.

[26] To a degree, companies like SpaceX are showing that increased redundancy may yield greater resiliency, which may offer alternatives to other aspects of DI. This point requires further research. For more background on SpaceX, see Passant Rabie, "SpaceX Launches 3,000th Starlink Satellite as Elon's Internet Constellation Continues to Grow," Gizmodo, August 10, 2022.

[27] Jeff Baumgartner, "Starlink Surpasses 400K Subscribers Worldwide," Light Reading, June 2, 2022.

[28] Viasat, "ViaSat-3 Is Designed to Unlock More Opportunity—for More of the World," webpage, undated.

[29] Sebastian Moss, "NEC to Build World's Highest Capacity Submarine Cable for Facebook, Shuttling 500 Tbps from U.S. to Europe," Data Center Dynamics, October 12, 2021.

[30] Mike Dano, "Starlink's Network Faces Significant Limitations, Analysts Find," Light Reading, September 23, 2020.

FIGURE 2.6

Satellite Networks and Supply Chains

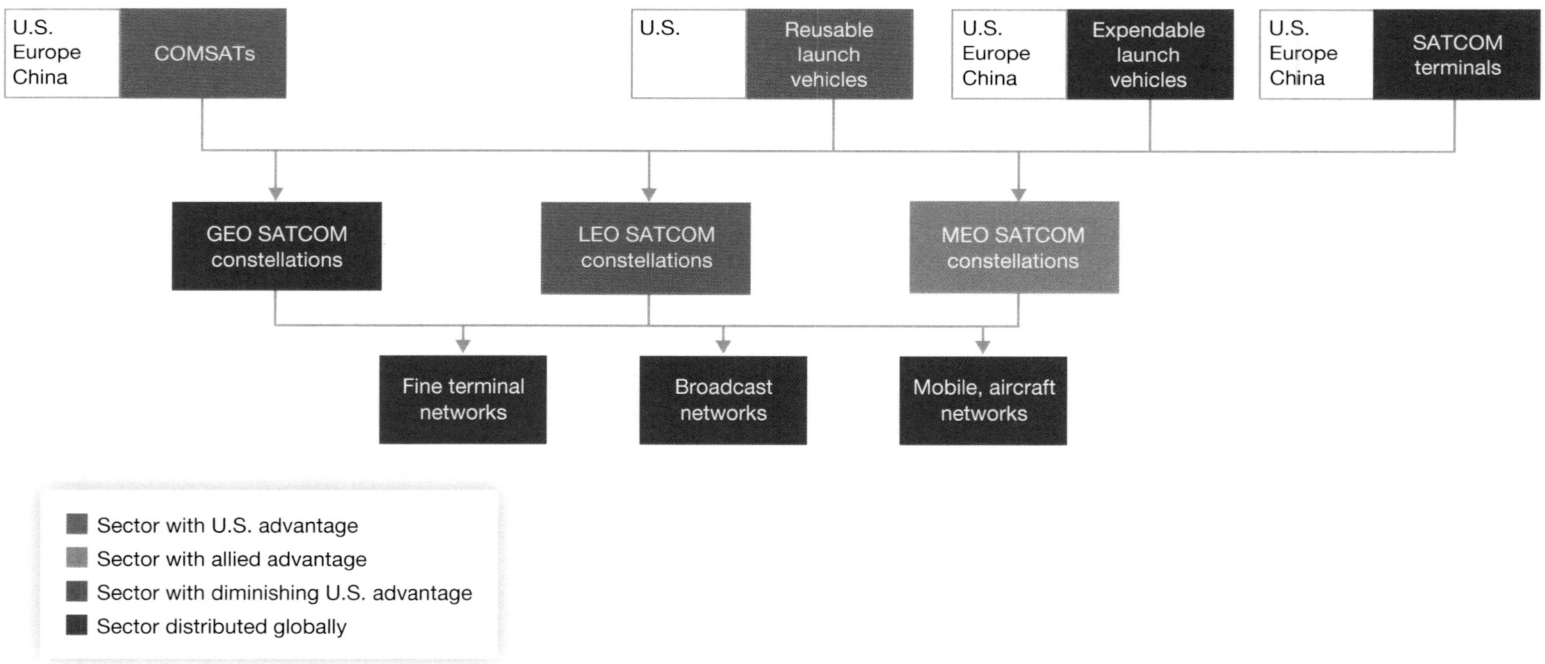

SOURCES: Authors' analysis of information from Brackup, Harting, and Gonzales, 2022; and Satellite Industry Association, 2022.

Network Components and Supply Chains

As indicated in Figure 2.6, the United States, Europe, and China are all manufacturers of GEO COMSATs. However, China does not export satellites to non-Chinese companies. Chinese companies play a much smaller role in the global COMSAT market, and most Chinese COMSATs provide communications services only in China.

U.S. and European companies currently lead the market for manufacturing the new generation of LEO COMSATs and deploying them in orbit. SpaceX makes its own Starlink satellites in the United States, and OneWeb makes its LEO COMSATs in Europe and in the United States. The Chinese government—using its state-owned aerospace companies and a new startup, GalaxySpace—has announced plans for large LEO COMSAT constellations of 1,000 and 13,000 satellites, respectively. GalaxySpace launched its first test satellites into orbit in March 2022, but no Chinese company has yet started providing communications services.[31] In this one part of the global DI, the United States is clearly ahead of China.

The two other essential ingredients for COMSAT networks are the launch vehicles needed to place the satellites in orbit and the SATCOM terminals used on the ground, ships, or aircraft. SATCOM terminals are made in the United States, Europe, and China. The U.S. and Chinese supply chains for COMSATs and SATCOM terminals have been largely decoupled since the late 1990s, because COMSATs and some SATCOM terminals have been deemed by the U.S. government to contain export-controlled technology. Six U.S. COMSATs were launched from China using Chinese launch vehicles, but a large fraction of these launches ended in failure. After the failed launch of Intelsat 7A in 1996, concerns grew that China had been able to obtain valuable know-how and technology by launching U.S. satellites. After the Intelsat 7A failure, U.S. export control policy regarding space technologies was strengthened even over the objections of the U.S. aerospace industry.[32]

Space is a strategic technology development area for China, and space launch vehicles are a key enabling technology for SATCOMs. In 2021, China launched more rockets into space than any other country. In 2021, there were 144 space launches, of which 133 were successful.[33] The two nations with the most launches were China and the United States. China executed 53 successful launches and the United States 51.[34] However, it should be noted that the United States launched more COMSATs into space than China in 2021, because of the large number of Starlink satellites that SpaceX launched in that year.

---

[31] Nick Wood, "China Enters the LEO Space Race," Telecoms, March 9, 2022.

[32] John Hoffner, "The Myth of 'ITAR-Free,'" Center for Strategic and International Studies, May 15, 2020.

[33] Eric Berger, "The World Just Set a Record for Sending the Most Rockets into Orbit," Ars Technica, January 3, 2022.

[34] Todd Harrison, Kaitlyn Johnson, Makena Young, Nicholas Wood, and Alyssa Goessler, "Space Threat Assessment 2022," Center for Strategic and International Studies, April 4, 2022; Deng Xiaoci and Fan Anqi, "China Scores 55 Orbital Launches in Super 2021, Topping U.S. to Become 1st in the World," Global Times, December 23, 2021.

It is also worth noting that as of 2022, SpaceX is the only provider of a commercial reusable launch vehicle, the Falcon 9 rocket, that can deliver satellites into orbit. Several Chinese companies are trying to develop reusable rockets; as of spring 2023, none has demonstrated the capability to launch a payload into space using a reusable launch vehicle. In the new technology of reusable launch vehicles, the United States is the clear leader, as indicated in Figure 2.6. The United States is also a technology leader in the development of both LEO and GEO COMSATs.

Foundational Elements and Technologies

The DI is based on two fundamental ingredients: microchips and technical standards. Microchips are essential for transmitting, receiving, and storing digital information. As each generation of more-advanced microchips is released and as these microchips are incorporated into the DI, the DI can process more information in less time.

Technical standards define how DI components transmit, process, and receive information. DI technical standards are incorporated into microchips and the software architectures for DI networks. DI technical standards also enable equipment from different vendors to be *interoperable*—i.e., they can interpret data received from other systems correctly. Different technical standards are used in each segment of the DI. However, some technical standards are common across the DI as described below, which enables information to flow seamlessly from satellites to terrestrial or wireless networks and through submarine cables.

Microchips

Microchips were invented in the United States and helped transform the U.S. economy by enabling the introduction of low-cost PCs, internet communications, and, later, countless other devices that generate and process digital information. However, over the past several decades, the manufacturing of microchips has transitioned offshore. In 2022, the majority of microchips used in the global economy were manufactured in Asia, and only 12 percent were still manufactured in the United States.[35] In 2019, 92 percent of the leading-edge microchips were made by just one company, TSMC in Taiwan, even though the United States retained a dominant market position in microchip software design applications and led in the production of many but not all types of semiconductor manufacturing equipment.[36]

China also depends on foreign suppliers for most of its microchips. The Chinese government has recognized the importance of microchips not only for its economy but also for its military because microchips are used in smart weapons, satellites, command and control systems,

---

[35] Deng and Fan, 2021.

[36] Antonio Varas, Raj Varadarajan, Ramiro Palma, Jimmy Goodrich, and Falan Yinug, "Strengthening the Global Semiconductor Supply Chain in an Uncertain Era," Boston Consulting Group, April 1, 2021.

sensors, and advanced military platforms. Importantly, China has an initiative, Made in China 2025, which includes a goal to become self-sufficient in producing leading-edge microchips.[37]

Figure 2.7 illustrates the major elements of the global microchip supply chain and the countries playing key roles in the supply chain. Specialized materials are required in microchip manufacturing, including high-purity silicon for silicon wafers for microchip substrates. Japan is a key supplier of high-purity silicon wafers. Rare gases are also required for microchip manufacturing. One essential ingredient is the inert gas neon, which is used for lasers in manufacturing processes. Ukraine supplies almost 50 percent of the neon used by the microchip industry.[38]

One essential part of the microchip supply chain is the high-tech manufacturing equipment used in the multistep manufacturing processes for microchips. U.S. companies play a leading role in the semiconductor manufacturing equipment market and offer some of the most-advanced pieces of equipment needed to make leading-edge microchips. However, it should be noted that U.S. companies do not make one piece of equipment essential for making the most-advanced microchips—extreme ultraviolet (EUV) lithography equipment. As of 2022, only one company makes EUV lithography equipment, Advanced Semiconductor Materials Lithography (ASML), located in the Netherlands. ASML EUV machines are essential for making microchips with feature sizes of 7 nanometers (nm) or less. ASML EUV equipment is essential for industry microchip foundries that have 7-nm or smaller process nodes.

Another essential part of the microchip supply chain is the tools needed to design microchips. Many of the microchip design tools capable of producing designs for leading-edge chips are provided by U.S. companies, as indicated by the green box in Figure 2.7.

Figure 2.7 indicates that the United States is in a strong position in the manufacturing equipment and design tool stages of the microchip supply chain. However, as noted above, the United States relies on key allies for access to leading-edge chip foundries. As discussed below, the world leaders in microchip foundries—i.e., those able to produce leading-edge microchips—are Taiwan and South Korea.[39]

China is in an even weaker position than the United States when it comes to leading-edge microchip foundries. China can only produce 14-nm microchips,[40] although unproven claims have surfaced that one Chinese foundry owner, Semiconductor Manufacturing International Corporation (SMIC), has been able to produce 7-nm chips with no defects using

---

[37] Karen M. Sutter, "'Made in China 2025' Industrial Policies: Issues for Congress," Congressional Research Service, IF10964, updated March 10, 2023.

[38] Alexandra Alper, "Exclusive: Russia's Attack on Ukraine Halts Half of World's Neon Output for Chips," Reuters, March 11, 2022. Neon is a by-product of some steel manufacturing processes.

[39] The manufacturing process for leading-edge chips can include over 400 individual steps and require the careful control of production process to prevent defects from occurring during manufacturing. The manufacturing processes include carefully guarded trade secrets of microchip foundry owners.

[40] Gonzales et al., 2022.

FIGURE 2.7

Microchip Supply Chains

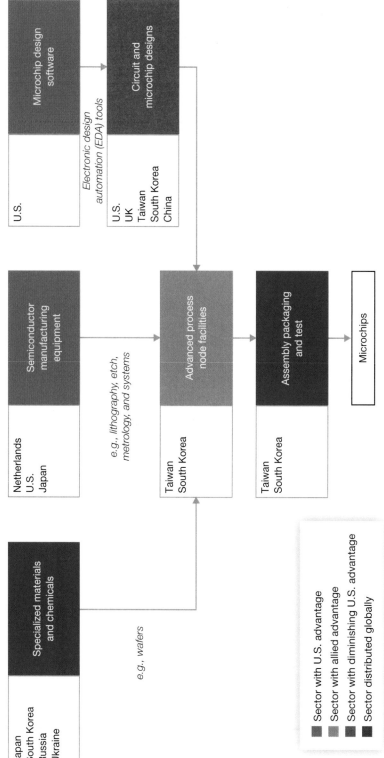

SOURCES: Authors' analysis of information from Satellite Industry Association, 2022; and Varas et al., 2021.

older-generation deep ultraviolet lithography equipment.[41] Key questions remain regarding the SMIC claims, such as the chip yields achieved by SMIC with its new 7-nm process and whether it is an economically viable foundry.

The last step in the microchip production process is packaging the chips into structures that protect the integrated circuits and provide the connections or pinouts so the chip can be connected to other chips on circuit boards. Most microchip packaging is done in Asia, and a growing share of this step is done in China. During this step, chips are also tested to verify they are authentic and that their circuits function according to their design specifications. Packaging factories are located in several countries, and so there is less concern that this step represents a critical vulnerable node in the microchip supply chain.

Figure 2.8 shows how the capabilities of the microchip foundries of Intel, Samsung, and TSMC have advanced over the past five years and how two companies predict they will advance in 2023. The figure shows that Intel, the one U.S. company that had until recently been a leader in microchip manufacturing, has fallen behind TSMC and Samsung. Intel has remained processing 10-nm microchips for more than three years. During this time, starting in 2019, Intel has made incremental improvements to its chips, but it has acknowledged that it has fallen behind its competitors.[42]

Figure 2.8 shows that the two leading microchip makers, TSMC and Samsung, were both producing 3-nm chips at volume in 2022. As of the third quarter of 2022, Intel was still producing microchips at 10 nm, although it had promised to introduce a 7-nm class chip by the end of 2022.

The U.S. government has grown increasingly concerned about the national security risks of relying on microchip foundries located in Taiwan and South Korea. These foundries are located in potential conflict zones. In conflict, or even during a crisis, these foundries could suffer damage or be taken offline for extended periods because of cyberattacks or sabotage. Virtually all leading-edge microchips (7 nm or below) come from TSMC and Samsung foundries in this region.[43] A conflict that disabled these foundries would have major implications for the U.S. economy and DoD supply chains. In response, the Biden administration proposed a set of initiatives to bring state-of-the-art microchip manufacturing back to the

---

[41] Anton Shilov, "SMIC Mass Produces 14nm Nodes, Advances to 5nm, 7nm," Tom's Hardware, September 16, 2022b.

[42] Characterizing semiconductor manufacturing production processes by a single number—the process node dimension—is an oversimplification of complex manufacturing processes and the complex geometry of microchips. Analysts generally agree that a 14-nm Intel chip is roughly equivalent to a 10-nm TSMC chip, that an Intel 10-nm process node chip is roughly equivalent to a 7-nm TSMC chip in terms of transistor dimensions, and that Intel microchips have greater transistor density at these levels (Anton Shivlov, "Intel's 10nm Node: Past, Present, and Future," *Electronic Engineering Times*, June 15, 2020; Ian Cutress, "Intel's Process Roadmap to 2025: with 4nm, 3nm, 20A and 18A?!" *AnandTech*, July 26, 2021). Nevertheless, even after making an adjustment that accounts for the higher transistor density of Intel's microchips, as of 2020, it was clear that Intel had fallen behind its two competitors in high-speed logic chips.

[43] Varas et al., 2021.

FIGURE 2.8

Process Nodes of Advanced Microchip Foundries

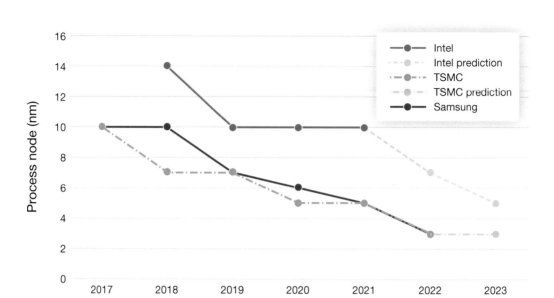

SOURCES: Features information from Cutress, 2021; Paul Alcorn, "Intel's 7nm Is Broken, Company Announces Delay Until 2022, 2023," Tom's Hardware, December 30, 2020; Anton Shilov, "Intel's 10nm Node: Past, Present, and Future," *Electronic Engineering Times*, June 15, 2020; Anton Shilov, "Intel Delays Mass Production of 10 nm CPUs to 2019," *AnandTech*, April 27, 2018; James Morra, "Samsung Foundry Delays 3-nm Node to 2022, 2-nm Due by 2025," Electronic Design, October 12, 2021; Samsung, "Samsung Electronics Expands Its Foundry Capacity with a New Production Line in Pyeongtaek, Korea," press release, May 21, 2020; Samsung, "Samsung Begins Chip Production Using 3nm Process Technology with GAA Architecture," press release, June 30, 2022; Samsung, "Samsung Showcases Its Latest Silicon Technologies for the Next Wave of Innovation at Annual Tech Day," press release, October 24, 2019; Samsung, "Samsung Electronics and Qualcomm Expand Foundry Cooperation on EUV Process Technology," press release, February 22, 2018; Shilov, 2022b; Areej Syed, "TSMC's 2nm Delay to 2026 May Allow Intel to Regain Process Leadership with Its 20A Wafers in 2024 or 2025," *Hardware Times*, June 18, 2022; TSMC, "3nm Technology," webpage, undated.

United States. In August 2022, Congress passed the Creating Helpful Incentives to Produce Semiconductors (CHIPS) and Science Act, which provides funding and incentives for U.S. and foreign firms to build microchip foundries in the United States.

Technical Standards

International technical standards are essential for communications networks so systems can be designed to interoperate with each other. Such standards contain many technical details and are produced by experts appointed to international standards development organizations (SDOs). Technically oriented SDOs use consensus-based or voting methods to determine the specifics of technical standards in a process in which companies offer proprietary technical information to the SDO for inclusion in a prospective standard. SDOs use deliberative processes and voting procedures to choose between differing technical approaches proposed by competing companies. As long as all parties act in good faith and vote to institute the best technical approach and approve the best technologies, DI technical standards will lead to a better and more secure DI in the future.

Some SDOs develop detailed technical standards, such as the 3GPP, which is responsible for developing technical standards for wireless cellular networks.[44] 3GPP uses voting rules to select technical standards. In this report, we will not go into the details about how technical standards are selected and how companies can offer their IP for inclusion in these standards.[45] However, it is worth noting that Chinese companies have been very active in proposing system essential patents (SEPs) and technical standards for future wireless networks to the 3GPP. Two Chinese 5G infrastructure providers, Huawei and ZTE, have proposed more SEPs, one of the first steps in defining a 3GPP technical standard, than any company from any other country.[46] Nevertheless, it appears that the technical quality of many Chinese-proposed 5G SEPs has been judged to be inferior than others proposed by other companies; thus, few Chinese SEP applications have been approved by 3GPP.[47] Another important SDO is the International Telecommunication Union (ITU), which recommends standards for telecommunications networks. The ITU also allocates frequencies for SATCOM networks and has become increasingly active in trying to set standards and influence governance over the internet.[48]

---

[44] 3GPP.2020, "5G Release 16," homepage, undated.

[45] See Gonzales et al. (2022) for more details.

[46] Danielle Pletka and Brett D. Schaefer, "Countering China's Growing Influence at the International Telecommunication Union," American Enterprise Institute, March 7, 2022.

[47] Gonzales et al., 2022.

[48] A Chinese national, Houlin Zhao, has been the head of the ITU since 2015. Since that time, the number of Chinese citizens employed at the ITU has increased, and Zhao has used his post to champion China's Belt and Road Initiative (BRI) and DSR Initiative (Pletka and Schaefer, 2022). In addition, Zhao has dismissed U.S. security concerns regarding Huawei's 5G network equipment:

> "Those preoccupations with Huawei's equipment, up to now there is no proof so far," Zhao said. "I would encourage Huawei to be given equal opportunities to bid for business, and during the operational process, if you find anything wrong, then you can charge them and accuse them. But if we don't have anything then to put them on the blacklist—I think this is not fair." (Pletka and Schaefer, 2022)

Under Zhao, the ITU has attempted to play a bigger role in internet governance and technical standards using the term *cyber sovereignty* that many observers believe is a set of internet standards that are intended to enable authoritarian governments to censor internet traffic and curtail free speech online (James Griffiths, *The Great Firewall of China: How to Build and Control an Alternative Version of the Internet*, Bloomsbury Publishing, 2019). Zhao stepped down at the end of 2022 (Tom Wheeler, "The Most Important Election You Never Heard Of," Brookings Institution, August 12, 2022), and the new ITU Secretary-General, Doreen Bogdan-Martin, started her term in January 2023 (U.S. Department of State, "Statement of Doreen Bogdan-Martin Upon Election to ITU Secretary-General," press release, September 29, 2022).

CHAPTER 3

Digital Infrastructure 2050 Futures

What could DI look like in 2050? What could an advantage or disadvantage in DI look like in alternative futures? Straight-line projections for how DI could evolve between now and 2050 assume one plausible path, but the interplay of different variables could create much different futures. In this chapter, we describe plausible futures for DI: a baseline future that assumes straight-line projections of current technology trends and three alternative futures that challenge core assumptions.

We generated these 2050 futures through an examination of elements shaping the DI competition (technology development, international use and adoption, access to alternatives) and the interplay among variables related to those elements. As discussed earlier, each element of the competition consists of a series of variables, such as economic performance and STEM workforce, which can be thought of as dials with different settings. We first determined the straight-line projections for these elements and variables based on the assessment of DI technologies in the previous chapter. We then turned the dials to create different settings that led to alternative future environments. What resulted was four DI futures, a baseline and three alternatives, which we have named as follows:

- Baseline Future: DI Battleground (diffuse and contested OAC with no real winners)
- Alternative Future 1: Cyber Fortress (China controls DI hardware landscape, but the United States maintains leadership in software)
- Alternative Future 2: Technological decoupling (bifurcated DI landscape, with separate U.S. and Chinese technology stacks and democratic and authoritarian regional DI blocs)
- Alternative Future 3: Rise of the Corporations (competition between U.S. tech giants and the Chinese Communist Party [CCP], with a limited U.S. government role).

The rest of this chapter describes these futures in greater detail and provides a discussion of key trends and variables.

Baseline Future: DI Battleground

By 2050, the United States has ceded some advantage in DI to China. Persistent challenges in shoring up domestic production for key DI technologies, issues in growing and sustaining a strong STEM workforce, and pervasive IP theft by People's Republic of China (PRC) entities compromising U.S. advanced technology stymied U.S. efforts to retain DI leadership. Nevertheless, U.S. companies still maintain a presence in key DI supply chains, in part because of U.S. policy limiting the transfer of advanced DI-related technologies to China and strengthened relationships with the commercial sector and U.S. allies and partners. OAC of the DI remains diffuse and competitive.

Meanwhile, China's long-term initiatives and ambitions paid off to some degree. Chinese investments in R&D were successful, and TSMC suffered serious missteps that cost them the advantage in semiconductors. China achieves rough parity with the West in advanced microchips.[1]

Chinese strategic initiatives, such as DSR, continued throughout the 2030s and 2040s and resulted in China's ability to compete for leadership in terrestrial fiber optics, switch and router markets, submarine cable networks, and cellular infrastructure. Both China and the United States continue to compete for leadership in the DI, with pockets of advantage for both in select areas but no clear leader across all.

Within terrestrial network markets, the United States and China continue to compete for leadership. Chinese SOEs have begun to own and operate many overseas terrestrial networks, which is not an approach that U.S. companies have actively pursued. Chinese SOEs dominate telecommunications markets in Southeast Asia (SEA), Africa, and Latin America in particular. U.S. and allied and partner companies remain dominant across Europe, North America, and parts of Asia.

As for submarine cables, U.S. big tech firms and PRC telecom SOEs own the greatest share of submarine cable networks, although telecoms and other companies from other countries co-own cables with Chinese or U.S. companies. Fewer cables have both U.S. and Chinese owners. Neither side has a dominant position in global submarine cable markets, and, in some regions, U.S. submarine cable networks compete with Chinese ones directly. Both the United States and China have maintained submarine cable supply chains, and neither side has a clear lead in technology, although the United States is dependent on Western allies for cable-laying ships. Also, PRC firms lead in cable laying and have majority OAC of landing sites in the regions where China dominates telecommunications markets (SEA, Africa, and Latin America).

As for wireless networks, by 2050, the United States has not regained a major presence in the cellular infrastructure market and instead relies on South Korean and European infrastructure providers. In comparison, Huawei is dominant in the cellular infrastructure market; after Huawei gains access to Chinese state-of-the-art microchips, it increases exports globally. Most of Europe and all NATO countries continue to follow U.S. policy and do not incorporate Huawei equipment into their wireless networks. China's SOEs focus on domi-

---

[1] Progress slows on both sides in the development of each successive generation of microchips because of the expense and growing technical difficulty in creating transistors with smaller feature sizes in microchips. Nevertheless, even in 2050, a new technology has yet to emerge to replace the capabilities of microchips needed to make ever more powerful mobile devices, computers, and high-bandwidth DI components.

nating markets again in select regions where it has seen success for other DI markets (SEA, Africa, and Latin America).

The United States retains a leadership role in space communications, driven by the new satellite networks offered by its commercial sector. U.S. COMSAT exports have grown significantly, while China lags two generations behind in SATCOM technology. However, both U.S. firms and Chinese SOEs are successful in deploying more-capable and competing mega-constellations in LEO. U.S. space networks are more capable, but they are more expensive than Chinese ones.

Furthermore, both the United States and China retain access to state-of-the-art microchips. China retains a strong influence over cellular standards bodies (e.g., 3GPP) and the ITU. The United States in 2050 has focused more attention on international SDOs, and so both the United States and China maintain influential roles in key SDOs governing the design of the DI. China is unable to dominate the ITU. The internet remains secure and based on open technical standards. In addition, U.S. satellite companies retain access to key frequency bands over much of the world, as do Chinese companies. U.S. and Chinese space companies, however, rely less on international standards and continue to develop and use their own technical standards, leading to noninteroperable satellite networks.

Key Assumptions

In this future, there is no clear advantage for the United States or China in DI, but each lead in select DI building blocks. Key assumptions in this future include China's ability to peacefully take over Taiwan, given their continued access to advanced microchips, and a strong position in the cellular infrastructure market in many but not all regions of the world. The United States and its allies in Europe and Asia have access and control over their DI infrastructure. However, China not only controls its own DI, but it has also expanded its control and access to the DI in several regions of the world, namely parts of Africa, Latin America, and SEA. The DI remains globally interconnected though its terrestrial, wireless, and submarine cable segments.

Alternative Future 1: U.S. Cyber Fortress

By 2050, the United States and China remain global economic and technology leaders, but where they lead within the DI differs. China has gained control of the global DI hardware market, except for space, through exploiting its dominance of the microchip manufacturing market. Key government initiatives within the United States to shore up a domestic chip manufacturing capability, such as the 2022 CHIPS and Science Act, ultimately failed to garner sufficient support in subsequent decades. U.S. chip makers fail to catch up to technology leaders in Asia. All leading-edge microchips are now made in China. Therefore, Chinese DI hardware components are used globally. As a result, U.S. reliance on Chinese hardware components grows over the decades. However, the United States is able to retain control of operating software systems core to the DI, such as iOS, Windows, Android, and the like. The

United States effectively creates a cyber fortress. What emerges is a codependency of sorts in 2050: The United States and its allies rely on China for state-of-the-art microchips, while U.S. companies license their software to Chinese and other companies that make computers, mobile phones, and, in some cases, DI components.

The United States recognizes China's advantage in controlling the hardware of the DI and the threat posed by their dominance of these markets. The United States attempts to prevent China from dominating U.S. DI and that of allied countries. As a result, the United States invests heavily in cybersecurity software and systems to find and identify malicious circuitry embedded within hardware and prevent it from compromising the operation of DI and computer components. In 2050, the United States is a leader in cybersecurity software services that can detect, isolate, and eliminate software-based cyber threats. The United States becomes the global leader in software- and firmware-based cybersecurity services, which reduces but does not eliminate cyber-based IP theft operations conducted by China.

The U.S. government coordinates its cybersecurity operations with U.S. tech giants and commercial companies. European allies and partners follow suit; the United States and Europe close their DIs using firewalls and advanced cybersecurity capabilities in an attempt to minimize the flow of information and intelligence from the West to China. The United States also pursues alternatives to microchips, such as biological and quantum processing, to lessen the long-term reliance on advanced microchips.

Meanwhile, China and Russia maintain strong and close ties but do not have integrated DIs. Both the United States and China maintain leadership roles in international SDOs. The United States remains active in DI-related SDOs and uses information gleaned from technical standards to understand how PRC-made chips should work.

The previous decade is marked by a tense intelligence and influence competition between the United States and China. China exploits its advantage in microchips to infiltrate and maintain a digital presence in national DIs globally, including the DI of the United States. The United States employs its cybersecurity capabilities to try and keep China out of its DI with mixed success. The United States has less success in infiltrating the Chinese DI and in maintaining insight into Chinese espionage, IP theft, and influence operations activities in the developing world.

By 2050, the United States has been able to build a software-based cyber fortress that minimizes the cyber capabilities of China inside the United States, but the United States is not immune to all cyber exploits embedded in Chinese microchips. Chinese malware and influence operations remain deeply embedded and integrated into the DI of DSR countries and even some U.S. allies. The U.S. and Chinese DIs remain partially open and connected as necessary, given the continued interdependencies of both countries.

Key Assumptions

The key assumptions in this future include China's growing advantage in DI hardware and the U.S. ability to maintain an advantage in software. Several key variables contribute to this future. First, a strong U.S. government partnership with commercial companies underpins the pursuit and development of the U.S. cyber fortress. Related, the United States must have access to a STEM workforce sufficient to support the cyber fortress. Both the United States and China pursue a systematic and centralized government approach to DI, and both countries maintain a domestic DI capacity and capability. Both countries have substantial, highly qualified STEM workforces, but both countries have workforce and other limitations that require them to specialize in either hardware or software. China, however, is a dominant player in DI globally, motivating the U.S. need to create a cyber fortress. U.S. allies in Europe and Asia adopt the exportable elements of the U.S. cyber fortress architecture, which curtails the IP theft operations of China in these economies. In this future, the DI remains a homogeneous and interoperable collection of networks based on international technical standards.

Alternative Future 2: Technological Decoupling

Decades of DI competition between the United States and China and sustained government efforts and investments to further DI OAC result in a technological decoupling for information and DI technologies by 2050. Two separate DI blocs emerge—a DI bloc led by the United States and its allies and partners and a DI bloc led by China—with little interoperability between the two, by design. International technical SDOs fracture into two separate organizations, except for the ITU, which has its standard-setting responsibilities curtailed in nearly all fields except for SATCOMS; where it remains responsible for the allocation of frequencies for these space systems.

In the run-up to 2050, China's digital surveillance export business and DSR efforts prove successful. Through these efforts, China outcompetes Western companies in terrestrial and wireless markets in Africa, South and Central Asia, and South America. However, because of growing privacy concerns, many countries in Europe and Asia choose smart city systems and wireless infrastructures from U.S. and European companies. The starkly different privacy protection capabilities of Western and Chinese DIs lead to further decoupling between these networks.

Meanwhile, some European telecommunications and space companies falter in the late 2020s and early 2030s and agree to join the new U.S.-led SDOs that agree to set new technical standards with embedded privacy protections for DI. These companies pivot to support U.S. companies and U.S. SDO–developed technical standards for DI. Western governments exclude China from these new international SDOs. These efforts result in a more integrated and more secure DI for participating Western countries.

In response, China forms its own competing DI SDOs. Other countries agree to participate but can offer little technology to the Chinese-led DI development effort. As a consequence, the declining Chinese STEM workforce becomes challenged to develop a DI with all of the capabilities of the Western DI and all the new and separate cyberattack tools needed to penetrate and compromise the Western DI without detection.

The number of submarine cables directly connecting the United States, NATO allied countries, Australia, and Japan to China fall dramatically, because no agreement can be reached on the standards used on these submarine cable networks and on cost-sharing arrangements. Two distinct and incompatible submarine cable networks develop globally, which prevents China from deploying many methods of IP theft and espionage against Western countries. A similar network decoupling occurs in SATCOMs, as it becomes illegal to export Western SATCOM terminals to China and other countries using Chinese DI. These changes also impact global businesses and trade, because communications become difficult and delayed between Western companies and suppliers in China. Chinese exports to many Western countries fall dramatically, while trade rises with countries in Asia that have chosen to connect to the Western DI, such as the Philippines and Vietnam.

Both the United States and China take deliberate government action supported by commercial activity to decouple integrated DI supply chains. Across submarine cable networks, terrestrial networks, wireless networks, and space infrastructure, PRC components are removed from the U.S. DI bloc, and, similarly, U.S. components are removed from the PRC DI bloc.

For microchips, two separate supply chains emerge; state-of-the-art microchips are developed for different technical standards and manufactured within each DI bloc.

Key Assumptions

In this future, the United States and China have regional advantages, but contested zones remain. It remains uncertain what DI advantage looks like in a completely decoupled world with two interoperable models for DI. Key variables associated with this future are the industrial base, STEM workforce, relationships with allies and partners, and overseas DI strategy. Both countries can maintain a sufficient STEM workforce to support their DI blocs with a large domestic capacity within their industrial base but with some reliance on third parties. Both countries also have a comprehensive strategy for DI and strong relationships with allies and respective partners. But global DI remains competitive, with regional DI blocs emerging and some contested zones.

Alternative Future 3: Rise of the Corporations

Failed government reforms and regulatory efforts, political stagnation, and domestic unrest have created conditions in which public trust in the U.S. government's ability to regulate economic activity and govern has declined. Meanwhile, U.S. tech giants have emerged as major centers of gravity exerting power and influence at scale—driving economic growth, strengthening the military, and controlling segments of the DI and the data transmitted, stored, and generated on DI networks. Tech giants, with their unique ownership and control of the DI and related information infrastructure, are the only organizations able to provide basic communications functions needed by the populace that were once provided by governments and government-regulated entities, such as traditional telecoms. In turn, the U.S. government relies more on tech giants in order to exert its limited control and influence.

U.S. tech giants, not governments, are the dominant players in the DI for the United States. The U.S. government plays a supporting role, but private companies drive the U.S. domestic and international agendas. Anti-trust legislation fails to rein in U.S. tech giants, which only grow and become more dominant. Data, owned and controlled by tech giants, become a source of economic value and power. This turn of events creates conditions in which tech giants have their own view and pursuit of national interests. Even more, tech giants begin to chart their own path for securing data, relying less on the U.S. government for protection and security and instead taking matters into their own hands. The U.S. government finds it challenging to exercise any control over DI.

As a result, U.S. tech giants compete with the Chinese government within the DI. China, in part threatened by the unwieldy power that U.S. tech giants have shown, plays a stronger hand internally. By 2050, the CCP has more success in exerting and solidifying state control over its tech giants, the DI, consumer data, social media, and information in general. The CCP, with the help of Chinese tech giants, exploits DI to conduct widespread surveillance of people inside of China and in DSR countries. The extensive monitoring and control by the CCP slows and weakens the technology innovation of Chinese tech giants. Chinese tech giants and PRC SOEs remain slow but competitive in select geographic areas (Africa, South and Central Asia, and South America) because of the subsidized low prices they offer and IP theft.

The shifting balance of power and the role of tech giants relative to companies have changed the U.S.-China competition for global DI. For the submarine cable market, U.S. tech giants compete globally against Chinese telecoms. In the terrestrial market, Amazon buys Cisco Systems to retain its competitiveness, but Huawei also remains a major player. Within the wireless market, Huawei fights for market share against U.S. and allied tech giants, while Google buys Ericsson and Samsung buys Nokia. Within the space sector, SpaceX competes with Kupier/Amazon, and Geely and GalaxySpace emerge as major competitors in SATCOMs for China. In terms of microchips, both the United States and China retain access to advanced microchips. The U.S. government's influence in international standards–setting bodies is minimized, creating a situation in which U.S. tech giants compete against the CCP in the deliberations of SDOs that set international technical standards.

Key Assumptions

In this future, the U.S. government loses influence and control in a way that begins to upend the international system and the traditional understanding of nation states. Data become the new source of economic power and influence, controlled and owned by major tech giants in the United States. These trends shape international dynamics and national interests. Within this system, U.S. tech giants retain a DI advantage but compete with the CCP as U.S. government influence is minimized.[2] Key variables contributing to this future are the govern-

---

[2] While government influence has been minimized, the government remains supportive of U.S. companies and U.S. interests writ large.

ment's relationships with tech giants and its approach toward economic policy. The United States retains a strong STEM workforce and large domestic capacity for DI within its industrial base, driven by the tech giants. CCP efforts to retain control over Chinese tech giants have hampered innovation, affecting China's STEM workforce and industrial base. U.S. tech giants have a dominant role in the global DI, but some regions are competitive with China's SOEs because of low prices and continued efforts to steal IP.

Summary of Relative Digital Infrastructure Advantage Across Futures

The futures described above portray different and complex landscapes for DI. Across all, China and the United States have varying degrees of advantage related to DI, in terms of their OAC of DI technologies or their access to trusted alternatives.

In the baseline future, DI remains contested in several areas. China has the advantage in submarine cables through OAC and in the wireless sector through both OAC and access to alternatives. The United States retains access to alternatives in the wireless sector but still is at a lesser advantage than China. Both compete for advantages in terrestrial, microchips, and technical standards.

As for alternative future 1 (U.S. cyber fortress), China has a dominant DI position, which motivated U.S. efforts to build a cyber fortress. This cyber fortress enables the United States to retain some access to alternatives through advanced cyber tools and measures. Advantage could be interpreted differently in this future depending on the weight placed on hardware versus software. The United States has an advantage in software, which China (and the rest of the world) needs to have for a fully functioning DI.

Alternative future 2 (technological decoupling) paints a picture of a divided global DI in which both the United States and China have separate and nonintegrated DIs. As a result, both maintain OAC and access to trusted alternatives through their DI blocs, but little advantage relative to one another. Compared with alternative future 1, the technological decoupling future posits a situation in which both the United States and China have almost full technological independence of each other; in alternative future 1, there is interdependence in DI hardware.

Finally, alternative future 3 (rise of the corporations) portrays a U.S. advantage but with the caveat that U.S. tech giants, not the U.S. government, shape the competition. The U.S. government has a minimized role in this future because of the significant power shown by U.S. tech giants and the inability of the government to control their growth. China has retained some OAC and access to alternatives but remains in competition for DI with U.S. tech giants. These futures warrant further analysis to understand relative advantages and disadvantages.

CHAPTER 4

Military Implications of Digital Infrastructure Futures

How could the global DI footprint affect future military operations? To answer this question, we used previous RAND research examining U.S. and Chinese approaches to current and future warfare to derive the military implications of 2050 DI futures.[1] The analysis also explored related questions: Does DI advantage—OAC and, in some cases, access to trusted alternatives—confer military advantages? What types of DI advantage might confer advantages for competition and conflict?

We emphasize several caveats. First, the literature review that underpins these implications leveraged existing U.S. and PLA strategy documents and research but does not reflect a comprehensive or definitive analysis of U.S. and Chinese military strategy or operations. Instead, it aimed to inform our speculation into how different DI futures might affect future warfare from U.S. and Chinese perspectives. Additionally, the military implications themselves are insights and considerations only and should not be viewed as definitive conclusions. The implications remain speculative in nature given the 2050 time frame explored, even though they are informed by previous research and analysis and subject-matter expertise. Lastly, the implications provide considerations for how to think about DI and how DoD might gain a better understanding of the U.S.-China DI competition.

U.S. and Chinese Military Activities for Competition and Conflict

The United States and China develop, sustain, modernize, and exercise military forces in support of national interests and objectives. How each military pursues its distinct set of objectives through a series of activities differs in some ways and remains similar in others. We note these activities in Table 4.1. For those activities noted in Table 4.1, DI could serve as both an enabler and a stand-alone capability to be leveraged for military advantage or influence. Before presenting these activities, we discuss U.S. and Chinese objectives.

---

[1] Brackup, Harting, and Gonzales, 2022.

U.S. military objectives remain consistent across competition and conflict, but the activities in support of these priorities vary per phase. The 2022 National Defense Strategy (NDS) fact sheet outlines the DoD's four primary priorities as follows:

- "defending the homeland, paced to the growing multi-domain threat posed by the PRC"
- "deterring strategic attacks against the United States, Allies, and partners"
- "deterring aggression, while being prepared to prevail in conflict when necessary" prioritizing the PRC challenge in the Indo-Pacific, then the Russia challenge in Europe
- "building a resilient Joint Force and defense ecosystem."[2]

The NDS further states that to achieve these objectives, DoD will pursue a series of activities focused on operating in multiple domains, improving alliances and partnerships, and using military power in concert with other aspects of national power to deter, compete, and operate effectively.[3]

China's national strategy, termed *national rejuvenation*, centers on China returning to a great-power status. The 2021 DoD China Military Power Report (CMPR) describes the aim of this strategy as achieving the "China Dream," defined as "a national aspiration to restore the PRC to a position of strength, prosperity, and leadership on the world stage."[4] Part of the goals and milestones underpinning the China Dream and strategy of national rejuvenation include modernizing the PLA and fielding a "world class" military.[5] Part of the Chinese efforts to achieve the China Dream and grow Beijing's global influence involves a series of overseas development efforts.[6]

Given this strategic context, Table 4.1 outlines related U.S. and Chinese military activities for competition and conflict. This understanding of objectives and activities informed our effort to assess the DI's relationship to these activities and how various DI futures might affect U.S. and Chinese military operations.

Competition

U.S. activities in competition largely consist of nonkinetic actions to improve partnerships with other countries, collect intelligence, and set the necessary conditions for conflict, such as gaining overseas access. More specifically, the United States conducts security cooperation efforts, such as partner training, bilateral and multilateral exercises, and military sales.[7] The

---

[2] DoD, "2022 National Defense Strategy," fact sheet, undated, p. 1.

[3] For more information, see DoD, undated, p.2, which discusses integrated deterrence, campaigning, and building enduring advantages.

[4] DoD, *Military and Security Developments Involving the People's Republic of China*, 2022, p. 1.

[5] DoD, 2022, pp. 3–4.

[6] DoD, 2022, p. 7.

[7] Joint Publication 3-20, *Security Cooperation*, U.S. Department of Defense, May 23, 2017.

TABLE 4.1

U.S. and Chinese Military Activities

Competition	Conflict
United States	
• Security cooperation • Strengthening allies and partnerships • Military-to-military exchanges • Overseas basing, access agreements • Ensuring free and open global commons • Intelligence collection • Military exercises (bilateral and multilateral) • Cyber operations • Information operations	• Distributed, joint, and multidomain operations • Projecting power quickly, flowing forces overseas to deter and defeat adversary aggression • Operating from overseas bases • Generating combat power from the homeland • Cyber operations • Intelligence collection • Information operations
China	
• BRI • DSR • Intelligence collection • Espionage, IP theft • Building relationships with other countries for basing and influence • Humanitarian assistance, disaster relief • Buildup of military infrastructure overseas • Cyber operations • Information operations • Proxy warfare • Territorial disputes (flying sorties, etc.)	• Informatization • Active defense • Intelligentization • Systems warfare • Target-centric warfare • A2/AD • Cyber operations • Information operations

SOURCES: Authors' analysis of information from Kathleen H. Hicks, Alice Hunt Friend, Joseph Federici, Hijab Shah, Megan Donahoe, Matthew Conklin, Asya Akca, Michael Matlaga, and Lindsey Sheppard, *By Other Means, Part I: Campaigning in the Gray Zone*, Center for Strategic and International Studies, July 2019; Headquarters, Department of the Army, *The Army in Military Competition*, Chief of Staff Paper No. 2, March 1, 2021; DoD, undated; DoD, *Summary of the 2018 National Defense Strategy of the United States of America: Sharpening the American Military's Competitive Edge*, January 2018; Zalmay Khalilzad and David A. Ochmanek, *Strategic Appraisal 1997: Strategy and Defense Planning for the 21st Century*, RAND Corporation, MR-826-AF, 1997; DoD, 2022; Edmund J. Burke, Kristen Gunness, Cortez A. Cooper III, and Mark Cozad, *People's Liberation Army Operational Concepts*, RAND Corporation, RR-A394-1, 2020; M. Taylor Fravel, *Active Defense: China's Military Strategy Since 1949*, Princeton University Press, 2019; Joel Wuthnow and M. Taylor Fravel, "China's Military Strategy for a 'New Era': Some Change, More Continuity, and Tantalizing Hints," *Journal of Strategic Studies*, 2022; Jeffrey Engstrom, *Systems Confrontation and System Destruction Warfare: How the Chinese People's Liberation Army Seeks to Wage Modern Warfare*, RAND Corporation, RR-1708-OSD, 2018; John Chen, Joe McReynolds, and Kieran Green, "The PLA Strategic Support Force: A 'Joint' Force for Information Operations," in Joel Wuthnow, Arthur S. Ding, Phillip C. Saunders, Andrew Scobell, and Andrew N. D. Yang, eds., *The PLA Beyond Borders: Chinese Military Operations in Regional and Global Context*, National Defense University Press, 2021; Mark Cozad, "Toward a More Joint, Combat-Ready PLA?" National Defense University Press, 2019; Kevin Pollpeter and Amanda Kerrigan, *The PLA and Intelligent Warfare: A Preliminary Analysis*, Center for Naval Analysis, 2021; Ronald O'Rourke, *Renewed Great Power Competition: Implications for Defense—Issues for Congress*, Congressional Research Service, R43838, updated August 3, 2021.

NOTE: A2/AD = anti-access and area denial.

United States also works with partner nations through diplomacy or military engagement to develop overseas basing and access agreements needed to support military operations beyond the U.S. homeland.[8] Additionally, in a competition environment, the U.S. Navy will periodically conduct freedom of navigation operations (FONOPs) to ensure a free and open mari-

---

[8] Stacie L. Pettyjohn and Jennifer Kavanagh, *Access Granted: Political Challenges to the U.S. Overseas Military Presence*, 1945–2014, RAND Corporation, RR-1339-AF, 2016.

time environment.[9] Lastly, the United States conducts intelligence collection throughout all military phases, including competition. This list likely omits important competition activities that the U.S. military performs; however, it is intended to be more illustrative than definitive or comprehensive.

Similarly, China's competition-phase military activities likely focus on setting the conditions for conflict and creating and sustaining areas of advantage vis-à-vis other actors. The 2021 CMPR discusses the direct linkages between how China views economic, security, and development interests: The BRI "expands the PRC's overseas development and security interests; Beijing has signaled this will drive the PRC toward expanding its overseas military footprint to protect those interests."[10] In this way, China's competition activities and others discussed below likely cannot be disentangled from its economic and development initiatives. Therefore, such efforts as BRI and DSR serve to grow Beijing's influence overseas, create infrastructure potentially useful for military activities, and create relationships with other countries.[11] Other Chinese competition-phase activities likely include intelligence collection and the espionage and theft of IP.[12] Lastly, additional nonkinetic activities that likely fall into a competition environment could include humanitarian assistance and disaster response and the continued buildup of overseas infrastructure, such as in the South China Sea.[13]

The United States likely continues to perform intelligence operations in crisis and, depending on the context, might pursue offensive cyber operations and information operations to either coerce another actor or to shape an outcome favorably to the United States. In some instances, the United States could exercise military power through a combat aircraft sortie or an increased forward deployed military presence. Chinese military activities in a crisis environment might include intelligence, cyber, and information operations, as well as proxy warfare.[14] Additionally, China might use a limited form of military hard power, such as flying combat aircraft sorties near a disputed territory.[15]

---

[9] DoD, "U.S. Department of Defense Freedom of Navigation (FON) Program," fact sheet, February 28, 2017.

[10] DoD, 2022, p. 8.

[11] Nadège Rolland, ed., *Securing the Belt and Road Initiative: China's Evolving Military Engagement Along the Silk Road*, National Bureau of Asian Research, September 2019.

[12] Christopher Wray, "Responding Effectively to the Chinese Economic Espionage Threat," Department of Justice China Initiative Conference, Center for Strategic and International Studies, February 6, 2020.

[13] Mike Yeo, "China Upgrades Military Infrastructure on South China Sea Islands, Report Claims," *Defense News*, February 9, 2017.

[14] For more information on Chinese gray zone activities, see Bonny Lin, Cristina L. Garafola, Bruce McClintock, Jonah Blank, Jeffrey W. Hornung, Karen Schwindt, Jennifer D. P. Moroney, Paul Orner, Dennis Borrman, Sarah W. Denton, and Jason Chambers, *Competition in the Gray Zone: Countering China's Coercion Against U.S. Allies and Partners in the Indo-Pacific*, RAND Corporation, RR-A594-1, June 2022.

[15] Lin et al., 2022, p. vii.

Conflict

Conflict refers to major combat operations that include kinetic and nonkinetic activities. In this phase, some activities from other phases remain constant, such as intelligence collection and cyber operations. However, new activities emerge depending on what the actor is trying to achieve and how each military envisions prosecuting warfare to achieve those objectives on behalf of the nation. Thus, the activities we included in Table 4.1 reflect U.S. and PLA visions for current and future warfare. For the United States, these activities appear more general, whereas Chinese activities align more closely with military concepts of operation. This distinction traces back to available information from both militaries, including historical data, on how they might fight a future war against a great power. As with the other phases, Table 4.1 and the above discussion provide a foundation for the more speculative military implications, but they are not a definitive or comprehensive account of U.S. and PRC conflict activities.

The United States envisions conducting warfare as distributed, joint, and multidomain operations with combat power generated from the homeland and overseas. Inherent to the U.S. way of war is the ability to quickly flow and sustain forces overseas. Furthermore, the U.S. military emphasizes the ability to conduct operations from overseas territories to deter and defeat adversary aggression.[16]

In the medium to long term, DoD envisions cross-domain effects as critical to success in large-scale combat operations. Currently referred to as multidomain operations, DoD envisions that future warfare will be more competitive in every domain for the United States. Additionally, DoD envisions operating as a more dispersed and distributed force given the growing threat posed by potential adversaries to fixed sites and locations. Therefore, to be successful, U.S. forces will have to integrate effects across domains to support timely decision-making, robust command and control, and responsive sensing and targeting for distributed forces in a high-threat environment.

China continues to focus on defending its sovereignty and national interests through a series of military modernization efforts and military concepts of operation highlighting the role of information in achieving victory in battle. Previous RAND research on Chinese operational concepts notes that the PLA stresses the role of information "both as a domain in which war occurs and as the central means to wage military conflict."[17] *Informatization*, or "informatized war," prioritizes the need for information dominance, in addition to air and maritime superiority, in a military campaign.[18]

These concepts illustrate the PLA's view of information dominance as providing a key operational advantage to exploit and as a means to undermine an adversary military advan-

---

[16] David A. Ochmanek, *Determining the Military Capabilities Most Needed to Counter China and Russia: A Strategy-Driven Approach*, RAND Corporation, PE-A1984-1, 2022.

[17] Burke et al., 2020, p. 5.

[18] Burke et al., 2020, p. 7.

tage.[19] China scholars note that "[t]here is consensus among PLA theorists that control over the information domain allows a combatant to dictate the pace, direction, and locale of conflict at the strategic, operational, and tactical levels."[20] Informatization also underpins the PLA's thinking on conducting joint operations; recent research on China highlights that "informatization is the essence of integrated joint operations, which rely on information networks to integrate and systematize operations designed to obtain information superiority."[21]

The Chinese military regards the systems that store, transmit, generate, and receive data as significant to effective military operations. Previous RAND research finds that the PLA conceptualizes modern warfare as a confrontation between operational systems, or systems warfare.[22] The PLA emphasizes that targeting the information elements of the operational system should be prioritized first, indicating the likely importance of DI to the execution of this theory of warfare.[23] Another relevant PLA operational concept, target-centric warfare, fits within the systems warfare construct, focusing on identifying and prosecuting key nodes within an operational system "to achieve decisive effects with minimal collateral damage."[24] Furthermore, intelligentization, touted as a future operational concept, likely leverages even more of the DI for military operations than systems warfare. Recent analysis from the Center for Naval Analysis notes that informatized warfare focuses on "systems architecture that links sensors, information, and people," whereas intelligent warfare emphasizes "AI [artificial intelligence] and autonomy."[25] How the PLA envisions intelligent warfare or intelligentization remains an open question, but early analysis from Chinese publications suggests the criticality of data and information systems.

Military Implications for Baseline Future

United States

The baseline future presents a complicated and challenging environment to navigate for competition and, to a lesser but still important degree, for conflict. Because the baseline has diffuse and interconnected supply chains, it becomes difficult to discern which actors might

---

[19] Burke et al., 2020, p. 7. Importantly, much of the PLA's thinking on the role of information dominance to winning wars comes from watching the U.S. military fight wars in the 1990s in the Gulf War and Kosovo War. For more information, see Cozad, 2019, p. 205.

[20] Chen, McReynolds, and Green, 2021, p. 153.

[21] Cozad, 2019, p. 208.

[22] Engstrom, 2018.

[23] For a more detailed account of systems warfare, systems-of-systems, and the role of information in these concepts, see Engstrom, 2018.

[24] Burke et al., 2020, p. 15.

[25] Pollpeter and Kerrigan, 2021, p. 6.

have advantages or OAC in each building block. Before delving into more specific military considerations, we provide an overview of the more general implications that emerge for the U.S. military in the baseline future:

- DI supply chain interdependence poses vulnerabilities in competition and crisis, particularly for nonkinetic operations.
- The U.S. military might retain inherent advantages in conflict if it has conventional supremacy.
- U.S. military posture might be more limited or operations might be more vulnerable in regions where PRC dominates DI.
- The U.S. military remains reliant on its allies and partners and the commercial sector for trustworthy DI.
- Because the U.S. military has continued access to microchips, it can develop advanced military capabilities.

Competition

While DI offers a key enabler to competition activities, it could also present significant vulnerabilities for activities that involve communicating with partners and operating from overseas territory where the DI could be compromised. Importantly, these implications will vary based on where the U.S. operates from, for how long, the type of mission, and the DI footprint in that territory.

The DI baseline future will affect how the U.S. military conducts longer-duration overseas military operations that might include a rotational or permanent force presence, multilateral or bilateral military exercises, any cyber or intelligence activities, and activities to engage partners and allies. These activities all involve either using an overseas DI for military operations or communicating with another actor via their DI. For cyber and intelligence operations, relative DI advantage—OAC of DI or access to trusted DI—becomes especially important for protecting U.S. national security information and networks, while on the other hand potentially enabling access to desired information. For more periodic, unilateral activities, such as maritime FONOPS using carrier strike groups, there might be fewer implications if the carrier strike group has access to SATCOM capabilities. In a peacetime environment, we assumed little to no degradation in a SATCOM capability.

Lastly, in the baseline future, China has grown its OAC and DI footprint in SEA, Africa, and South America. The United States' ability to engage countries, whether on a diplomatic or more operational level, could be more limited in the baseline future for two reasons. First, given China's control of DI in these regions, Beijing might wield more influence and be able to deter countries from engaging with the United States. Second, because countries in this region will communicate using Chinese-made infrastructure, the U.S. military might deem it too risky to discuss more-sensitive issues, or any issues not necessarily public in nature. As a result, the United States can still work with countries of interest in these regions, but the nature of those interactions might be more limited because of DI considerations. Or, alternatively, the United States might need to bring its own DI to support operations.

Conflict

In conflict, DI can serve as an important enabler, a key vulnerability, or a stand-alone capability, depending on the DI landscape and types of operations. DI enables military communications; command and control; and the generation, storing, and transmitting of any data or information across systems, units, and individuals. Vulnerabilities in the U.S. military DI can provide access points for the adversary to gain an advantage. Conversely, U.S. advantages in global and domestic DI can provide the capability to gain an advantage vis-à-vis the adversary by accessing, and potentially disrupting or destroying, systems of interest. The nature of U.S. military operational concepts and broader military strategy in 2050 remains a key uncertainty; however, we drew implications for more general aspects of U.S. warfighting tied to projecting and sustaining power overseas, joint and multidomain operations, and operations in the cyber and information spaces. Therefore, the key military implications for U.S. operations in conflict in the 2050 baseline future surround where and how the U.S. military operates.

If the United States must operate for a duration that requires the military to use overseas DI (non-U.S.) in a country where China has a strong DI footprint, U.S. military commanders will likely need to determine whether and how the U.S. operates. Furthermore, for kinetic activities, the DI landscape will affect where those types of missions take place but might not alter the entire mission set even in a degraded environment. However, for nonkinetic activities tied to intelligence, cyber, and information operations, DI can serve as a capability with such advantages as early warning and the ability to disrupt the adversary's information and military communications.

China

As noted above, the diffuse nature of the baseline future poses challenges for China in competition and conflict as to where and how it conducts operations. Below are the general implications that emerge for the PLA in the baseline future:

- The PLA's ability to realize info-centric warfare at scale might be challenged by continued DI competition.
- China's control of DI is globally challenged, which might hamper its ability to achieve power and influence in some regions.
- China has a regional advantage for Indo-Pacific operations, with opportunities for potential overseas operations in select areas of responsibility (AoRs) (e.g., SEA, Africa, and South America).
- PLA relies on successful military-civil fusion for access to DI microchips, which enable the development of advanced military capabilities.

Competition

In competition, DI underpins activities for building relationships with and influencing other countries and intelligence, while providing a key enabler for other operations, such as humanitarian assistance and disaster relief. For building relationships with other countries, DI appears central to BRI. Specifically, the DSR lays out a clear approach to engaging overseas actors through the selling, installment, and operation of DI technology. Additionally, countries that use Chinese-made DI offer promising areas for the PLA to operate from, if desired. In this way, many of China's activities in competition will set the conditions for conflict. The baseline future posits that China has DI OAC in SEA, Africa, and South America, improving its ability, if desired, to build up military infrastructure in those regions.

On the other hand, the baseline future presents both opportunities and areas of vulnerability for intelligence, cyber, and information operations. Given that China appears to use cyber capabilities for espionage activities, DI OAC can enable that to a degree. Furthermore, because DI fundamentally deals with the generation, storing, and transmission of data, OAC of DI technology facilitates intelligence and information operations. However, the interdependent supply chains posited in the baseline future might make it more difficult to discern DI vulnerabilities from advantages because the technology consists of many layers, components, and owners (see Chapter 2).

Conflict

The PLA's emphasis on targeting information systems and gaining information dominance to defeat an adversary suggests that the overall DI landscape might have important implications for China in a conflict. In the baseline future, for conflicts in Asia, Africa, or South America, the PLA will likely have the necessary DI to support military operations. However, given China's lack of advantage in space, PLA expeditionary operations might be limited to AoRs where China has a large DI footprint. However, given that the United States has access to alternative trustworthy DI and a separate set of microchips and has superiority in space, the PLA's ability to realize such concepts as informatization and intelligentization will likely give neither country an advantage. Therefore, kinetic operations might be favored to target information nodes and systems of interest. Importantly, the PLA's access to DI for military operations relies on successful military-civil fusion. Additionally, the PLA's access to microchips enables it to develop and use advanced military capabilities.

Military Implications for Alternative Future 1: Cyber Fortress

In this future, China owns and controls DI hardware globally through control of microchips, but the U.S. has developed advanced software and cyber capabilities to retain access to trustworthy DI.

United States

To protect U.S. DI devices and networks, the United States closed off its internet and broader DI using firewalls. Only European countries using similar firewalls and cyber capabilities fall within this U.S.-led closed DI. Lacking a traditionally open internet and communications infrastructure has several implications for U.S. military operations and competition. We summarize these implications below:

- Military communications and operations globally become more challenging.
- The sheer number of Chinese DI globally presents a targeting and data-sharing challenge.
- Rotational or permanent presence overseas will likely be limited to Europe and some parts of Asia.
- A reliance on robust software-based cyber capabilities will be necessary to compete effectively and access DI.
- The U.S. government will be reliant on the commercial sector to develop software and cyber.
- The development of advanced military capabilities might lag because of increased cyber needs.

First, military communications on a global scale become significantly more complex in this future given the need to add advanced software and cyber capabilities to counteract compromised DI hardware. Similarly, the United States will likely be limited in who it can interact with given interoperability issues. The United States' closed and tailored DI might not be interoperable with other countries' DIs. Therefore, any activities to work with other non-European partners might need to either be more in-person and rely less on such functions as email or other digital formats. Additionally, the sharing of information with countries that use DI incompatible with U.S. DI—i.e., do not have the requisite advanced software capabilities—or those countries that do not fall within the United States' closed DI, might take a more physical form as well, such as printed materials. Finding and targeting key Chinese information nodes, whether for competition or conflict purposes, will likely be more difficult given the ubiquity of Chinese DI globally. Mapping Chinese DI and understanding where the critical nodes or centers exist might not be feasible. Furthermore, the ubiquity of Chinese DI offers significant resilience for Beijing.

Specifically for conflict, the United States might rely more on generating combat power from the homeland where it has trustworthy DI rather than projecting power forward given DI limitations. However, the U.S. military might create a new mission set that involves developing and bringing DI forward to establish expeditionary networks leveraging a DI that has incorporated the new software and cyber capabilities. Again, it will limit how the United States might interact with host nation partners not already within the United States' closed DI. Another important implication from this future involves the rate of technological development for the United States. Because the United States needs to develop software to sanitize Chinese-made microchips, this requirement will likely slow the rate of development of advanced capabilities—military and civilian—that rely on state-of-the-art semiconductors. This could create significant limitations over the long term for the relative U.S.-China balance of military hard power. Lastly, this alternative future makes the U.S. military entirely reliant on the commercial sector for access to DI, stressing the need for amicable and robust public-private relationships.

China

China's dominance in the microchips market enables it to control the majority of DI hardware globally and likely use DI both as an enabler and capability to create military advantage. We summarize some of these implications below:

- The control of hardware improves the PLA's ability to conduct info-centric warfare.
- A DI advantage, particularly in Asia, might limit U.S. ability to operate in the Indo-Pacific.
- An increased ability to project global power and potential means results in the ability to deter or deny U.S. access to parts of world.
- China is given a greater opportunity for the offensive use of DI given the global ubiquity of Chinese DI and fewer vulnerabilities for an adversary to exploit it.

First, the PLA discusses targeting information systems to create advantage and defeat an adversary while also using information dominance and networked operations to win battles. Control of DI hardware at scale could allow the PLA to realize this vision of warfare non-kinetically, depending on its degree of global OAC and its ability to use that OAC to create decisive nonkinetic effects when desired. On the other hand, the PRC's control of microchips, and thus DI hardware, gives it an advantage in protecting its own networks and information. The absence of a convoluted supply chain will likely make it easier for China to identify vulnerabilities, while making it more difficult for adversaries to create and exploit them. Lastly, the PRC's control of DI hardware on a more-global scale might deter the United States from operating in regions or territories dominated by Chinese DI, at least for longer-duration missions. If the United States cannot communicate or share information digitally because of interoperability issues or risk of compromise concerns, the United States might be denied from operating in certain areas.

Military Implications for Alternative Future 2: Technological Decoupling

Technological decoupling in 2050 would result in two heterogenous and distinct DIs with little interoperability. China would lead one DI model, and the United States would lead the other. This future presents many implications for posture, military interoperability, and basic communications and information-sharing between countries.

United States

A true DI technological decoupling poses significant military implications for how the United States would interact with partners, allies, other countries of interests, and other societies more broadly. We provide a snapshot of these higher-level implications below:

- U.S. allies and partners enable the United States' ability to conduct expeditionary operations and project power globally.
- Operations in the Indo-Pacific become more challenging and vulnerable with more-limited basing options.
- Contested DI zones remain in the Arctic, Middle East, North Africa, and potentially elsewhere.
- The U.S. military likely realizes its vision of warfare based on power projection, but it is limited to certain AoRs.
- Global intelligence operations become more limited

Given interoperability challenges between the DI models, such competition activities as strengthening relationships with allies and partners will be limited to countries within the U.S. DI model (i.e., countries that use U.S.-made DI). More starkly, any bilateral or multilateral interaction will largely be confined to countries that fall within the U.S. model. In a conflict environment, the United States, if political accesses allow, could potentially execute longer-duration missions from territories within the U.S. model more easily than it does today. Because the United States largely leads one of the DI models, it might have more insight into the supply chains and knowledge of vulnerabilities that could be mitigated. In a diffuse DI landscape, identifying and mitigating vulnerabilities in the DI becomes very challenging. On the other hand, leveraging DI as an enabler and capability for nonkinetic activities, such as intelligence collection and cyber operations, becomes somewhat uncertain in a decoupled world. Without globally interconnected networks, access to systems and information remotely might become very difficult. Similarly, any U.S. personnel operating in a territory using Chinese DI might have significant challenges with communicating to a command node located in the United States or in an allied or partner country. Lastly, contested zones might persist where parts of a country use one version of DI and another part uses a different model. We hypothesized that these regions might include the Arctic, parts of the Middle East and North Africa, and potentially others. U.S. operations in these regions might be limited to standoff activities given sensitivities tied to the compromise of data or information. Alternatively, the United States might simply begin to bring the entire DI with them overseas.

A technologically decoupled world might also introduce new missions for the United States to ensure that the military can operate as needed to achieve its objectives. These missions might include developing a mobile DI that has all the necessary hardware and software components. The military could then transport the entire DI stack—all necessary components to create a network—to the AoRs to enable U.S. military activities.

China

China's ability to conduct military operations in a technologically decoupled world might be geographically limited and challenge Beijing's ability to conduct intelligence and espionage activities. We provide a snapshot of the potential military implications for China in a decoupled future below:

- The Chinese military finds it more difficult to achieve information dominance, which could lead the PLA to rely on kinetic measures to target information nodes of interest.
- The ability to conduct extended or long-range operations beyond the Chinese DI bloc are limited.
- Military capabilities could be risked if the U.S.-led DI bloc innovates faster and the PRC cannot access technology or IP to develop a comparable Chinese version.
- Global intelligence operations are more limited.

As noted earlier, the PLA views information dominance, and the ability to target key information nodes, as critical to winning battles. Therefore, in a decoupled world where access to those systems in a nonkinetic way becomes extremely challenging, kinetic options likely appear more promising. Similar to the United States' ability to operate overseas, China's ability to operate from overseas territories for longer-duration missions will likely hinge on the type of DI used in the host country. However, from a competition standpoint, China's ability to engage regions of interest (South America, South and Central Asia, and Africa) has increased because of the bifurcation in technology models. The United States chose not to compete in these regions, facilitating China's ability to create and foster influence.

Military Implications for Alternative Future 3: Rise of the Corporations

In the last alternative future, it is unclear what warfare would look like in a world where states compete with tech giants for power and influence. While the United States has a DI advantage in this future, it also remains unclear where the advantage lies for military operations.

United States

Although military advantage or disadvantage largely cannot be inferred from this future, we present some potential insights into the military projection of power below:

- The transition of power from government to tech giants could potentially completely upend traditional DoD structures: Who would control the military? What would the military be used for?
- The relationship between DoD and tech giants is unclear and might be stronger in some DI segments than in others.
- Tech giants control networks and network-related technology, and they can degrade the network as needed.

More questions than insights emerge in this future for U.S. military operations. Namely, if the private sector comes to dominate major power centers traditionally run by the government, what role would military power play for the United States?

China

Beijing faces many challenges militarily in this future, particularly for capability modernization and force employment. We summarize some of these high-level considerations below:

- China likely faces a more-limited ability to innovate because of weakened tech giants.
- Military operations might be more limited and regionally focused, but they maintain a structure to employ forces and capabilities.
- Unbridled U.S. tech giants might be able to degrade DI used by the PRC for military operations.
- China could potentially be unable to realize algorithmic warfare before the United States.

First, this future posits that the government has weakened the country's tech giants, likely degrading their ability to contribute to broader technological innovation that could underpin military capabilities. Second, given the U.S. advantage throughout the DI, and its larger DI footprint globally, the PLA might be more limited in terms of where they can operate overseas. Lastly, China might be slow to realize their vision of algorithmic warfare because of the lack of a strong tech industry within China. Algorithmic warfare would likely involve a robust workforce and set of organizations to develop the software and capabilities to implement intelligentization and intelligent warfare. However, a major uncertainty of this future remains how the CCP would compete with U.S. companies compared with the U.S. government.

Summary of Military Implications

The global DI footprint might affect how forces operate, where forces operate, and when forces operate. First, the DI futures suggest that U.S. and Chinese forces might conduct military operations differently depending on the DI landscape in the AoR and the type of mission. For example, power projection and longer-duration overseas presence become more difficult in any future in which China has DI dominance. In this way, DI could affect how forces operate in the future.

Furthermore, the futures also indicate that DI footprints appear to have a significant impact on military posture. Network boundaries introduce new types of borders between actors; most futures result in regional blocs determined by DI borders. The United States might choose not to have a permanent or rotational presence in territories that use Chinese DI given concerns over compromised information and communications. On the other hand, the United States can still generate combat power from the homeland and overseas, but when this can occur might depend on the global DI laydown. For example, a lack of an overseas permanent or at least rotational presence might affect how quickly the United States can flow forces to an overseas location. Related to posture considerations, sustained overseas operations will likely look very different depending on the DI landscape. Furthermore, the types of relationships with allies and partners might be limited based on DI laydown (e.g., weapon interoperability will likely be limited in more-decoupled futures; the ability to share information might be limited). As a result, the overall DI landscape could create more bifurcation between who the United States operates with and coordinates with militarily and who the United States cannot work with because of DI limitations (exacerbated by political and policy factors).

DI futures might also create a demand for new mission sets, new capabilities, and perhaps a new organization of forces. The futures show that DI could become like jet fuel in some decoupled futures—something the military needs to take with them when operating overseas. For example, a new mission could be to transport and operate a mobile U.S. DI. The futures also suggest that intelligence and cyber activities might be challenged depending on the degree of U.S.-Chinese DI interdependence or decoupling. Other mission sets or capabilities could also emerge in response to a lack of traditional cyber accesses and intelligence operations in an interconnected digital world. In a decoupled world, countries across the globe will not be interconnected like they are today, particularly the United States and China. This disconnection will make cyber and intelligence operations inherently more difficult, introducing the potential need for new missions and capabilities.

Insights on Digital Infrastructure

DI appears significant for future military operations and long-term strategic competition both as an enabler and as a capability. The global DI footprint, and how it evolves, might affect how forces operate, where forces operate, and when forces operate. And, by doing so, DI could confer key military advantages or disadvantages for different actors and shape outcomes for competition and conflict. Perhaps most significantly for the U.S.-China competition, DI advantage could potentially provide China an asymmetric means to erode U.S. military advantage over the long term. We provide an initial set of insights gleaned from this analysis, with a caveat noting their speculative nature. We include them to contribute to the broader understanding of the DI competition and inform future DI research.

Ceding DI OAC to an adversary could potentially provide an asymmetric means to erode U.S. military advantage over the long term. While the U.S. military does not appear unable to conduct military operations without DI OAC (i.e., the United States can likely conduct strike missions without control of the DI), access to information and systems could be leveraged for advantage to degrade or exploit U.S. capabilities, operations, and other functions, such as planning. Additionally, DI OAC offers the ability to gain access to systems and networks that could be leveraged in a conflict environment for destructive purposes. It could take decades to gain a superiority in DI that would confer such advantages, with the vulnerabilities unknown until a conflict takes place. On the other hand, gaining DI OAC before another competitor offers key advantages for national and military power.

Structural factors will shape how DI evolves and, ultimately, the DI footprint. STEM workforce (human capital more generally), economic performance and policy, industrial capacity, government approach to DI, and relationships with allies and partners all underpin what the DI will look like in the future. Therefore, these factors offer important areas of focus when thinking about how to shape the future of the DI in ways favorable to the United States.

DI could serve as an enabler or a capability for military operations. In other words, military operations might rely on DI to execute missions or leverage DI as the means to create decisive effects. For example, sending sensing information to a shooter will likely require some aspect of the DI, particularly if using a command-and-control system. Additionally, to share intelligence information with another country, the DI will be required. On the other hand, DI as a capability, what we term *digital presence*, could be the means through which a military degrades or destroys a military target in conflict. The distinction between DI as an enabler versus DI as a capability offers insights into varying degrees of OAC implications. If military forces, or specific military mission sets, rely on DI as an enabler but not as a capability, forces might be able to afford differing degrees of DI OAC and yet still be successful. However, if military forces use DI as a capability, DI OAC might be the only means to create a digital presence that can be leveraged to access military targets or other systems of interest.

For the United States, DI appears most significant for competition activities and setting the conditions for conflict, if it were to emerge. As a result, DI OAC matters greatly for competition, and activities that determine DI OAC ultimately dictate the DI footprint for conflict. For conflict, the ability to conduct kinetic operations will likely continue regardless of DI OAC, assuming the United States has alternative methods for basic communications and transmitting data. In this way, DI OAC might facilitate kinetic activities and create various advantages, but the lack of DI OAC likely does not restrict the military from conducting strike missions. In a conflict environment, imposing costs will always matter, and how an actor creates costs on the enemy might differ depending on a variety of factors. Furthermore, the United States has little to no ability to shape the DI footprint once a conflict has broken out, particularly if with China. Therefore, activities in competition to shape the DI footprint become increasingly important. DI offers greater advantages in competition to work with allies and partners, create influence, and conduct intelligence operations. Without DI OAC in competition, these activities, particularly security cooperation, might become very difficult.

Considerations for the Department of Defense

This analytic effort sought to understand how the DI might evolve and the military implications for different DI futures. This research showed the linkage between DI and military operations and highlighted the growing importance of accounting for DI when analyzing long-term military competition. An actor's DI OAC or access to alternatives can create advantages or introduce vulnerabilities in military competition and conflict. As a result, efforts to assess long-term military competition should account for

- the DI footprint relative to areas and regions of focus
- DI comparative advantages and vulnerabilities.

This effort built on previous DI research and identified the following areas where further research is needed:

- understanding how China characterizes and frames the DI
- identifying what to measure and what metrics are appropriate with respect to DI in a comparative assessment
- understanding how a technological decoupling between the United States and China could occur and what different phases or levels of technological decoupling would look like
- developing conflict scenarios for select DI futures and a related operational assessment of how DI would be used in conflict.

Variables and Settings for Alternative Futures

For each alternative future developed, we used a matrix as an analytic tool to identify and explore different variables and settings. This appendix provides the matrix used for each future to show the variables and settings for each DI future in greater detail. Each matrix lists the DI issues and variables as rows and plausible settings (i.e., a range of plausible outcomes for these variables) as columns—the content is consistent across each future (see Chapter 2 for a more detailed discussion of these issues and variables and why they matter for DI). For each future, we set different combinations of plausible settings to first define a straight-line projection of how DI could evolve (to establish a baseline) and then to explore non–straight-line projections that were still plausible and highly consequential, even if less likely. For the baseline, we leveraged our understanding of the DI and consulted with other SMEs to make plausible assumptions for how the future might evolve along straight-line projections. The color coding in the matrix indicates how the dials were turned for the settings to create that future, with red indicating China's future status for a particular variable, blue indicating the United States' future status for a particular variable, and orange indicating the future status of both countries for a particular future. Below each of the matrices, we explain in greater detail some of the assumptions associated with the settings selected.

Baseline Future

Table A.1 shows a breakdown of the issues and variables involved in the baseline future. We assumed that China's high-quality STEM workforce will decline over time while the U.S. STEM workforce will remain stable. In addition, we assumed that the United States could continue to rely on partners and allies for key DI components and infrastructure. In terms of the industrial base, given current trends, we assumed that by 2050 China will acquire the capability to make state-of-the-art microchips and become self-sufficient in DI technology. In comparison, the United States will have a large domestic capacity but will continue to rely on trusted third parties for DI components and infrastructure (i.e., wireless infrastructure). We also assumed that DSR initiatives will continue and will give China the ability to con-

TABLE A.1
Baseline Future

Element	Variable	Setting A	Setting B	Setting C	Setting D
Tech development	STEM workforce	Strong	Sufficient	Declining	Insufficient
	Industrial base	Self-sufficient; DI technology protected, secure	Some reliance on trusted 3rd parties; large domestic capacity	Mostly reliant on 3rd parties; limited domestic capacity	Reliance on untrusted 3rd parties
	Government approach	Systematic/centralized	Siloed	Nonexistent by choice	Diminished government role/authority
	Economic policy	Laissez faire/free market	Centralized/directed	Mixed—some markets closed	Open to allies, closed to adversary
	Economic performance	Global leader	Growing	Declining	Poor
International use/adoption	Overseas DI strategy	Comprehensive national-level DI strategy and resources	Some offensive (funding), some defensive measures	Defensive-only policy (no use of adversary infrastructure)	Nonexistent
	Relationships with allies and partners	Strong	Strained	Limited	Weak
	Overseas OAC of DI	Global	Many regions	Own and allied countries	Only in own country
	Overall OAC OF DI	Dominant/monopoly	Diffuse/competitive	Limited	None; reliant on others
Access to alternatives	Access to alternatives	Yes; trusted	Yes, but untrustworthy	Limited	None

NOTE: Red shading = key variables for China; light blue shading = key variables for United States; orange shading = key variables for both the United States and China.

tinue to install DI in the developing world.[1] In comparison, we assumed that U.S. policy for overseas DI investments would remain unchanged absent major changes in the International Monetary Fund and World Bank. However, the United States will be able to work with its network of allies and partners to prevent the installation of untrustworthy Chinese DI infrastructure in these countries. Finally, the United States will retain access to DI alternatives because of its dominant position in space and because SATCOMs can serve as an alternative to other parts of the DI.

Alternative Future 1: Cyber Fortress

Table A.2 shows a breakdown of the issues and variables involved for the cyber fortress alternative future. Both the United States and China must rely on untrusted third parties for key DI components (China is reliant on software from the United States, whereas the United States is reliant on China for microchips.) However, we assumed that the United States would maintain a strong STEM workforce and associated defense industrial base to support its cyber fortress and might also limit the export of software source code, design tools, and cybersecurity tools to China.

Alternative Future 2: Technological Decoupling

Table A.3 shows a breakdown of the issues and variables involved in the technological decoupling alternative future. Both the United States and China almost completely decouple and become self-sufficient in advanced technology. The United States will share its technology with partners and allies, particularly those strong in producing trustworthy DI infrastructure in Europe and Asia.

Alternative Future 3: Rise of the Corporations

Table A.4 shows a breakdown of the issues and variables involved in the rise of corporations alternative future. We assumed that U.S. tech giants could grow over decades and develop, potentially through foreign R&D and production capacity in allied countries, an unparalleled high-quality STEM workforce. We also assumed that U.S. tech giants will develop their own cybersecurity architectures to limit and prevent systemic or widespread use of untrustworthy systems and components.

---

[1] Although we acknowledge that China is currently reconsidering certain BRI investments because of large financial losses encountered in some countries (e.g., Sri Lanka, Pakistan) (Lingling Wei, "China Reins in Its Belt and Road Program, $1 Trillion Later," *Wall Street Journal*, September 26, 2022).

TABLE A.2
Cyber Fortress Future

Element	Variable	Setting A	Setting B	Setting C	Setting D
Tech development	STEM workforce	Strong	Sufficient	Declining	Insufficient
	Industrial base	Self-sufficient; DI technology protected, secure	Some reliance on untrusted 3rd parties; large domestic capacity	Mostly reliant on 3rd parties; limited domestic capacity	Reliance on untrusted 3rd parties
	Government approach	Systematic/centralized	Siloed	Nonexistent by choice	Diminished government role/authority
	Economic policy	Laissez faire/free market	Centralized/directed	Mixed—some markets closed	Open to allies, closed to adversary
	Economic performance	Global leader	Growing	Declining	Poor
International use/adoption	Overseas DI strategy	Comprehensive national-level DI strategy and resources	Some offensive (funding), some defensive measures	Defensive-only policy (no use of adversary infrastructure)	Nonexistent
	Relationships with allies and partners	Strong	Strained	Limited	Weak
	Overseas OAC of DI	Global	Many regions	Own and allied countries	None
	Overall OAC of DI	Dominant/monopoly	Diffuse/competitive	Limited	None; reliant on others
Access to alternatives	Access to alternatives	Yes; trusted	Yes; untrustworthy	Limited	None

TABLE A.3

Technological Decoupling Future

Element	Variable	Setting A	Setting B	Setting C	Setting D
Tech development	STEM workforce	Strong	Sufficient	Declining	Insufficient
	Industrial base	Self-sufficient; DI technology protected, secure	Some reliance on untrusted 3rd parties; large domestic capacity	Mostly reliant on 3rd parties; limited domestic capacity	Reliance on untrusted 3rd parties
	Government approach	Systematic/centralized	Siloed	Nonexistent by choice	Diminished government role/authority
	Economic policy	Laissez faire/free market	Centralized/directed	Mixed—some markets closed	Open to allies, closed to adversary
	Economic performance	Global leader	Growing	Declining	Poor
International use/adoption	Overseas DI strategy	Comprehensive national-level DI strategy and resources	Some offensive (funding), some defensive measures	Defensive-only policy (no use of adversary infrastructure)	Nonexistent
	Relationships with allies and partners	Strong	Strained	Limited	Weak
	Overseas OAC of DI	Global	Many regions	Own and allied countries	None
	Overall OAC of DI	Dominant/monopoly	Diffuse/competitive	Limited	None; reliant on others
Access to alternatives	Access to alternatives	Yes; trusted	Yes; untrustworthy	Limited	None

TABLE A.4

Rise of the Corporations Future

Element	Variable	Setting A	Setting B	Setting C	Setting D
Tech development	STEM workforce	Strong	Sufficient	Declining	Insufficient
	Industrial base	Self-sufficient; DI technology protected, secure	Some reliance on untrusted 3rd parties; large domestic capacity	Mostly reliant on 3rd parties; limited domestic capacity	Reliance on untrusted 3rd parties
	Government approach	Systematic/centralized	Siloed	Nonexistent by choice	Diminished government role/authority
	Economic policy	Laissez faire/free market	Centralized/directed	Mixed—some markets closed	Open to allies, closed to adversary
	Economic performance	Global leader	Growing	Declining	Poor
International use/adoption	Overseas DI strategy	Comprehensive national-level DI strategy and resources	Some offensive (funding), some defensive measures	Defensive-only policy (no use of adversary infrastructure)	Nonexistent
	Relationships with allies and partners	Strong	Strained	Limited	Weak
	Overseas OAC of DI	Global	Many regions	Own and allied countries	None
Access to alternatives	Overall OAC of DI	Dominant/monopoly	Diffuse/competitive	Limited	None; reliant on others
	Access to alternatives	Yes; trusted	Yes; untrustworthy	Limited	None

Abbreviations

3GPP	3rd Generation Partnership Project
5G	fifth generation
AoR	area of responsibility
BRI	Belt and Road Initiative
CCP	Chinese Communist Party
CMPR	China Military Power Report
COI	community of interest
COMSAT	communications satellite
DI	digital infrastructure
DoD	U.S. Department of Defense
DSR	Digital Silk Road
EUV	extreme ultraviolet
FONOP	freedom of navigation operation
GEO	geosynchronous orbit
IoT	Internet of Things
IP	intellectual property
ISP	internet service provider
ITU	International Telecommunication Union
LAN	local area network
LEO	low earth orbit
MAN	metropolitan area network
MEO	medium earth orbit
NATO	North Atlantic Treaty Organization
NDS	National Defense Strategy
OAC	ownership, access, and control
PC	personal computer
PLA	People's Liberation Army
PRC	People's Republic of China
R&D	research and development
RAN	radio access network
SATCOM	satellite communications
SDO	standards development organization
SEA	Southeast Asia
SME	subject-matter expert

SOE	state-owned enterprise
SS7	Signaling System 7
STEM	science, technology, engineering, and mathematics
TSMC	Taiwan Semiconductor Corporation
WAN	wide area network

Bibliography

3GPP.2020, "5G Release 16," homepage, undated. As of May 4, 2023:
https://www.3gpp.org/release-16

Absolute Reports, "Wireless Router Market Size by 2022–2028 Key Players, Regional Segmentation, Types, Applications, Growth, Shares, Revenue, Opportunities, Challenges, Drivers, Trends," press release, April 29, 2022.

Alcorn, Paul, "Intel's 7nm Is Broken, Company Announces Delay Until 2022, 2023," Tom's Hardware, December 30, 2020.

Alper, Alexandra, "Exclusive: Russia's Attack on Ukraine Halts Half of World's Neon Output for Chips," Reuters, March 11, 2022.

Bateman, Jon, *U.S.-China Technological "Decoupling": A Strategy and Policy Framework*, Carnegie Endowment for International Peace, 2022.

Baumgartner, Jeff, "Starlink Surpasses 400K Subscribers Worldwide," Light Reading, June 2, 2022.

Berger, Eric, "The World Just Set a Record for Sending the Most Rockets into Orbit," *Ars Technica*, January 3, 2022.

Bonds, Timothy M., James Bonomo, Daniel Gonzales, C. Richard Neu, Samuel Absher, Edward Parker, Spencer Pfeifer, Jennifer Brookes, Julia Brackup, Jordan Wilcox, David R. Frelinger, and Anita Szafran, *America's 5G Era: Gaining Competitive Advantages While Securing the Country and Its People*, RAND Corporation, PE-A435-1, 2021. As of August 16, 2023:
https://www.rand.org/pubs/perspectives/PEA435-1.html

Brackup, Julia, Sarah Harting, and Daniel Gonzales, *Digital Infrastructure and Digital Presence: A Framework for Assessing the Impact on Future Military Competition and Conflict*, RAND Corporation, RR-A877-1, 2022. As of May 4, 2023:
https://www.rand.org/pubs/research_reports/RRA877-1.html

Brake, Doug, "Submarine Cables: Critical Infrastructure for Global Communications," Information Technology and Innovation Foundation, April 19, 2019.

Burke, Edmund J., Kristen Gunness, Cortez A. Cooper III, and Mark Cozad, *People's Liberation Army Operational Concepts*, RAND Corporation, RR-A394-1, 2020. As of July 20, 2023:
https://www.rand.org/pubs/research_reports/RRA394-1.html

Burnett, Douglas R., "Submarine Cable Security and International Law," *International Law Studies*, Vol. 97, No. 1, 2021.

Chen, John, Joe McReynolds, and Kieran Green, "The PLA Strategic Support Force: A 'Joint' Force for Information Operations," in Joel Wuthnow, Arthur S. Ding, Phillip C. Saunders, Andrew Scobell, and Andrew N. D. Yang, eds., *The PLA Beyond Borders: Chinese Military Operations in Regional and Global Context,* National Defense University Press, 2021.

Cozad, Mark, "Toward a More Joint, Combat-Ready PLA?" National Defense University Press, 2019.

Cutress, Ian, "Intel's Process Roadmap to 2025: With 4nm, 3nm, 20A and 18A?!" *AnandTech*, July 26, 2021.

Dano, Mike, "Starlink's Network Faces Significant Limitations, Analysts Find," Light Reading, September 23, 2020.

Deng Xiaoci and Fan Anqi, "China Scores 55 Orbital Launches in Super 2021, Topping U.S. to Become 1st in the World," *Global Times*, December 23, 2021.

Dewar, James A., *Assumption-Based Planning: A Tool for Reducing Avoidable Surprises*, Cambridge University Press, 2002.

DoD—*See* U.S. Department of Defense.

Engstrom, Jeffrey, *Systems Confrontation and System Destruction Warfare: How the Chinese People's Liberation Army Seeks to Wage Modern Warfare*, RAND Corporation, RR-1708-OSD, 2018. As of July 20, 2023:
https://www.rand.org/pubs/research_reports/RR1708.html

Expert Market Research, "Global Ethernet Adapter Market Report and Forecast 2022–2027," webpage, undated. As of September 1, 2022:
https://www.expertmarketresearch.com/reports/ethernet-adapter-market

Frankiewicz, Becky, and Tomas Chamorro-Premuzic, "Digital Transformation Is About Talent, Not Technology," *Harvard Business Review*, May 6, 2020.

Fravel, M. Taylor, *Active Defense: China's Military Strategy Since 1949*, Princeton University Press, 2019.

Glanz, James, and Thomas Nilsen, "A Deep-Diving Sub. A Deadly Fire. And Russia's Secret Undersea Agenda," *New York Times*, April 20, 2020.

Gonzales, Daniel, Julia Brackup, Spencer Pfiefer, and Timothy Bonds, *Securing 5G: A Way Forward in the U.S. and China Security Competition*, RAND Corporation, RR-A435-4, 2022. As of June 20, 2023:
https://www.rand.org/pubs/research_reports/RRA435-4.html

Griffiths, James, *The Great Firewall of China: How to Build and Control an Alternative Version of the Internet*, Bloomsbury Publishing, 2019.

Guenni, "German Wind Turbines, the War in Ukraine and the Broken Satellite Communication," Born's Tech and Windows World, March 1, 2022.

Hardy, Stephen, "Hengtong to Buy Huawei Marine Networks," Lightwave, November 4, 2019.

Harrison, Todd, Kaitlyn Johnson, Makena Young, Nicholas Wood, and Alyssa Goessler, "Space Threat Assessment 2022," Center for Strategic and International Studies, April 4, 2022.

Headquarters, Department of the Army, *The Army in Military Competition*, Chief of Staff Paper No. 2, March 1, 2021.

Heuer, Richards J., Jr., and Randolph H. Pherson, *Structured Analytic Techniques for Intelligence Analysis*, CQ Press, 2011.

Hicks, Kathleen H., Alice Hunt Friend, Joseph Federici, Hijab Shah, Megan Donahoe, Matthew Conklin, Asya Akca, Michael Matlaga, and Lindsey Sheppard, *By Other Means, Part I: Campaigning in the Gray Zone*, Center for Strategic and International Studies, July 2019.

Hillman, Jonathan E., *The Digital Silk Road: China's Quest to Wire the World and Win the Future*, Harper Collins Publishers, 2021.

Hoffner, John, "The Myth of 'ITAR-Free,'" Center for Strategic and International Studies, May 15, 2020.

International Data Corporation, "IDC's Worldwide Quarterly Ethernet Switch and Router Trackers Show Strong Growth in Fourth Quarter of 2021," press release, March 10, 2022.

Ip, Greg, "China's Rise Drives a U.S. Experiment in Industrial Policy," *Wall Street Journal*, March 10, 2021.

Joint Publication 3-20, *Security Cooperation*, U.S. Department of Defense, May 23, 2017.

Khalilzad, Zalmay, and David A. Ochmanek, eds., *Strategic Appraisal 1997: Strategy and Defense Planning for the 21st Century*, RAND Corporation, MR-826-AF, 1997. As of September 9, 2022: https://www.rand.org/pubs/monograph_reports/MR826.html

Kurlantzick, Joshua, "Assessing China's Digital Silk Road Initiative: A Transformative Approach to Technology Financing or a Danger to Freedoms?" Council on Foreign Relations blog, December 18, 2020.

Lempert, Robert J., Steven W. Popper, and Steven C. Bankes, *Shaping the Next One Hundred Years: New Methods for Quantitative, Long-Term Policy Analysis*, RAND Corporation, MR-1626-RPC, 2003. As of July 20, 2023: https://www.rand.org/pubs/monograph_reports/MR1626.html

Lewis, James Andrew, *Mapping the National Security Industrial Base: Policy Shaping Issues*, Center for Strategic and International Studies, May 2021.

Lin, Bonny, Cristina L. Garafola, Bruce McClintock, Jonah Blank, Jeffrey W. Hornung, Karen Schwindt, Jennifer D. P. Moroney, Paul Orner, Dennis Borrman, Sarah W. Denton, and Jason Chambers, *Competition in the Gray Zone: Countering China's Coercion Against U.S. Allies and Partners in the Indo-Pacific*, RAND Corporation, RR-A594-1, 2022. As of July 20, 2023: https://www.rand.org/pubs/research_reports/RRA594-1.html

Miller, Christopher, Mark Scott, and Bryan Bender, "UkraineX: How Elon Musk's Space Satellites Changed the War on the Ground," *Politico*, June 9, 2022.

Morgan, Emma, "Thousands in France Lose Internet in Suspected Russian Cyberattack," *The Connexion*, March 4, 2022.

Morgan, Ryan, "China Must Be Prepared to Destroy Elon Musk's Starlink System, Chinese Researchers Say," American Military News, May 25, 2022.

Morra, James, "Samsung Foundry Delays 3-nm Node to 2022, 2-nm Due by 2025," Electronic Design, October 12, 2021.

Moss, Sebastian, "NEC to Build World's Highest Capacity Submarine Cable for Facebook, Shuttling 500Tbps from U.S. to Europe," Data Center Dynamics, October 12, 2021.

Neustadt, Richard E., and Ernest R. May, *Thinking in Time: The Uses of History for Decision-Makers*, The Free Press, 1986.

Novico, Trish, "5 Largest Fiber Optic Companies in the World," *Insider Monkey*, December 3, 2020.

Ochmanek, David A., *Determining the Military Capabilities Most Needed to Counter China and Russia: A Strategy-Driven Approach*, RAND Corporation, PE-A1984-1, June 2022. As of July 20, 2023: https://www.rand.org/pubs/perspectives/PEA1984-1.html

"Optical Transceiver Market—Growth, Trends, COVID-19 Impact, and Forecasts 2023–2028," Mordor Intelligence, undated.

O'Rourke, Ronald, *Renewed Great Power Competition: Implications for Defense—Issues for Congress,* Congressional Research Service, R43838, updated August 3, 2021.

Page, Jeremy, Kate O'Keeffe, and Rob Taylor, "America's Undersea Battle with China for Control of the Global Internet Grid," *Wall Street Journal*, March 12, 2019.

Pearson, James, Raphael Satter, Christopher Bing, and Joel Schectman, "Exclusive: U.S. Spy Agency Probes Sabotage of Satellite Internet During Russian Invasion, Sources Say," Reuters, March 11, 2022.

Pettyjohn, Stacie L., and Jennifer Kavanagh, *Access Granted: Political Challenges to the U.S. Overseas Military Presence*, 1945–2014, RAND Corporation, RR-1339-AF, 2016. As of August 16, 2023:
https://www.rand.org/pubs/research_reports/RR1339.html

Pletka, Danielle, and Brett D. Schaefer, "Countering China's Growing Influence at the International Telecommunication Union," American Enterprise Institute, March 7, 2022.

Pollpeter, Kevin, and Amanda Kerrigan, *The PLA and Intelligent Warfare: A Preliminary Analysis,* Center for Naval Analysis, 2021.

Pottinger, Matt, and David Feith, "The Most Powerful Data Broker in the World Is Winning the War Against the U.S.," *New York Times*, November 30, 2021.

Rabie, Passant, "SpaceX Launches 3,000th Starlink Satellite as Elon's Internet Constellation Continues to Grow," Gizmodo, August 10, 2022.

Reim, Garrett, "Musk Deepens Support for Ukraine, Despite Russian Anger," Aviation Week Intelligence Network, March 8, 2022.

Reports and Data, *High-Speed Optical Transceiver Market*, 2022.

Research and Markets, "Global Optical Transceiver Market (2022 to 2027)—Featuring Amphenol, Broadcom and Infinera Among Others," press release, June 30, 2022.

Robuck, Mike, "Report: High-Speed Data Center Ethernet Adapter Market Tops $1B for the First Time," Fierce Telecom, May 20, 2020.

Rolland, Nadège, ed., *Securing the Belt and Road Initiative: China's Evolving Military Engagement Along the Silk Road*, National Bureau of Asian Research, September 2019.

Samsung, "Samsung Electronics and Qualcomm Expand Foundry Cooperation on EUV Process Technology," press release, February 22, 2018.

Samsung, "Samsung Showcases Its Latest Silicon Technologies for the Next Wave of Innovation at Annual Tech Day," press release, October 24, 2019.

Samsung, "Samsung Electronics Expands Its Foundry Capacity with a New Production Line in Pyeongtaek, Korea," press release, May 21, 2020.

Samsung, "Samsung Begins Chip Production Using 3nm Process Technology with GAA Architecture," press release, June 30, 2022.

Satellite Industry Association, *State of the Satellite Industry Report*, June 2022.

Shilov, Anton, "Intel Delays Mass Production of 10 nm CPUs to 2019," *AnandTech*, April 27, 2018.

Shilov, Anton, "Intel's 10nm Node: Past, Present, and Future," *Electronic Engineering Times*, June 15, 2020.

Shilov, Anton, "TSMC Roadmap Update: N3E in 2024, N2 in 2026, Major Changes Incoming," *AnandTech*, April 22, 2022a.

Shilov, Anton, "SMIC Mass Produces 14nm Nodes, Advances to 5nm, 7nm," Tom's Hardware, September 16, 2022b.

Silverberg, Elliot, and Eleanor Hughes, "Semiconductors: The Skills Shortage," The Lowy Institute, September 15, 2021.

Sutter, Karen M., "'Made in China 2025' Industrial Policies: Issues for Congress," Congressional Research Service, IF10964, updated March 10, 2023.

Syed, Areej, "TSMC's 2nm Delay to 2026 May Allow Intel to Regain Process Leadership with Its 20A Wafers in 2024 or 2025," *Hardware Times*, June 18, 2022.

Taiwan Semiconductor Manufacturing Company, "3nm Technology," webpage, undated. As of May 4, 2023:
https://www.tsmc.com/english/dedicatedFoundry/technology/logic/l_3nm

Tannenbaum, Andrew, *Computer Networks*, 4th ed., Prentiss Hall, 2003.

TelecomLead, "Cisco Replaces Huawei in Service Provider Router and Switch Market," March 11, 2022.

TeleGeography, "Submarine Cable Map," dataset, 2023. As of July 20, 2023:
https://www.submarinecablemap.com/ready-for-service/2023

"Top 5 Vendors in the Global Submarine Fiber Cable Market from 2017 to 2021," Business Wire, October 30, 2017.

"Top Fiber Optics Suppliers and Manufacturers in the USA," Thomasnet, undated.

TSMC—*See* Taiwan Semiconductor Manufacturing Company.

U.S. Department of Defense, "2022 National Defense Strategy," fact sheet, undated.

U.S. Department of Defense, "U.S. Department of Defense Freedom of Navigation (FON) Program," fact sheet, February 28, 2017.

U.S. Department of Defense, *Summary of the 2018 National Defense Strategy of the United States of America: Sharpening the American Military's Competitive Edge*, January 2018.

U.S. Department of Defense, *Military and Security Developments Involving the People's Republic of China*, 2022.

U.S. Department of Justice, "Team Telecom Recommends That the FCC Deny Pacific Light Cable Network System's Hong Kong Undersea Cable Connection to the United States," press release, June 17, 2020.

U.S. Department of State, "Statement of Doreen Bogdan-Martin Upon Election to ITU Secretary-General," press release, September 29, 2022.

Varas, Antonio, Raj Varadarajan, Ramiro Palma, Jimmy Goodrich, and Falan Yinug, "Strengthening the Global Semiconductor Supply Chain in an Uncertain Era," Boston Consulting Group, April 1, 2021.

Viasat, "ViaSat-3 Is Designed to Unlock More Opportunity—for More of the World," webpage, undated. As of May 4, 2023:
https://www.viasat.com/space-innovation/satellite-fleet/viasat-3/

Watts, Barry, *Analytic Criteria for Judgments, Decisions, and Assessments,* Center for Strategic and Budgetary Assessments, 2017.

Wei, Lingling, "China Reins in Its Belt and Road Program, $1 Trillion Later," *Wall Street Journal*, September 26, 2022.

Wheeler, Tom, "The Most Important Election You Never Heard Of," Brookings Institution, August 12, 2022.

Wieland, Ken, "Router/Switch Market Back to Pre-Pandemic Levels—Dell'Oro," Light Reading, June 1, 2021.

Winkler, Jonathan Reed, "Telecommunications in World War I," *Proceedings of the American Philosophical Society*, Vol. 159, No. 2, June 2015a.

Winkler, Jonathan Reed, "Silencing the Enemy: Cable-Cutting in the Spanish-American War," War on the Rocks, November 6, 2015b.

Wood, Nick, "China Enters the LEO Space Race," Telecoms, March 9, 2022.

Wray, Christopher, "Responding Effectively to the Chinese Economic Espionage Threat," Department of Justice China Initiative Conference, Center for Strategic and International Studies, February 6, 2020.

Wuthnow, Joel, and M. Taylor Fravel, "China's Military Strategy for a 'New Era': Some Change, More Continuity, and Tantalizing Hints," *Journal of Strategic Studies,* 2022.

Yang, Stephanie, "Chip Makers Contend for Talent as Industry Faces Labor Shortage," *Wall Street Journal*, January 2, 2022.

Yeo, Mike, "China Upgrades Military Infrastructure on South China Sea Islands, Report Claims," *Defense News*, February 9, 2017.

Zilberman, Alan, and Lindsey Ice, "Why Computer Occupations Are Behind Strong STEM Employment Growth in the 2019–29 Decade," *Beyond the Numbers: Employment & Unemployment*, Vol. 10, No. 1, January 2021.